T0254404

Cambridge Elements ≡

Elements in Physics beyond the Standard Model with Atomic and Molecular Systems
edited by
David Cassidy
University College London (UCL)
Rouven Essig
Stony Brook University
Jesús Pérez-Ríos
Stony Brook University

THE ROLE OF SYMMETRY IN THE DEVELOPMENT OF THE STANDARD MODEL

Sherwin T. Love
Purdue University

Shaftesbury Road, Cambridge CB2 8EA, United Kingdom

One Liberty Plaza, 20th Floor, New York, NY 10006, USA

477 Williamstown Road, Port Melbourne, VIC 3207, Australia

314–321, 3rd Floor, Plot 3, Splendor Forum, Jasola District Centre,
New Delhi – 110025, India

103 Penang Road, #05–06/07, Visioncrest Commercial, Singapore 238467

Cambridge University Press is part of Cambridge University Press & Assessment,
a department of the University of Cambridge.

We share the University's mission to contribute to society through the pursuit of
education, learning and research at the highest international levels of excellence.

www.cambridge.org
Information on this title: www.cambridge.org/9781009478601

DOI: 10.1017/9781009238427

First published 2024

A catalogue record for this publication is available from the British Library.

ISBN 978-1-009-47860-1 Hardback
ISBN 978-1-009-23845-8 Paperback
ISSN 2754-4621 (online)
ISSN 2754-4613 (print)

Cambridge University Press & Assessment has no responsibility for the persistence
or accuracy of URLs for external or third-party internet websites referred to in this
publication and does not guarantee that any content on such websites is, or will
remain, accurate or appropriate.

The Role of Symmetry in the Development of the Standard Model

Elements in Physics beyond the Standard Model with Atomic
and Molecular Systems

DOI: 10.1017/9781009238427
First published online: January 2024

Sherwin T. Love
Purdue University

Author for correspondence: Sherwin T. Love, loves@purdue.edu

Abstract: Symmetry and its various realizations have played a pivotal role in the development of the extremely well-tested Standard Model of the strong, weak and electromagnetic interactions. In this Element, the author traces the development of the model through the interplay of the different symmetries realized in the various components of the model as well as in other sub-fields of physics.

Keywords: symmetry, standard model of particle physics, Higgs mechanism, QED, QCD

ISBNs: 9781009478601 (HB), 9781009238458 (PB), 9781009238427 (OC)
ISSNs: 2754-4621 (online), 2754-4613 (print)

Contents

1 Introduction

In July 2012, the CMS and Atlas collaborations working at the Large Hadron Collider (LHC) at CERN in Geneva, Switzerland jointly announced [1],[2] the detection of a scalar boson of mass roughly 125 GeV decaying into two photons and exhibiting the properties of the Higgs boson, the particle remnant of the electroweak symmetry-breaking mechanism of the Standard Model (SM) of particle physics. For their theoretical work describing this mechanism, P. Higgs and F. Englert were jointly awarded the 2013 Nobel Prize in Physics. This prize was really all about the role of symmetry, its form and its realization, in the electroweak interactions. Symmetry has also played a crucial role in the development of the theory of strong interactions, Quantum Chromodynamics (QCD), the other component of the Standard Model. In fact, one might argue that the deeper understanding of symmetry in many ways underlies much of the physics of the past century. In this Element we discuss the role symmetry played leading to the construction of the SM. The treatment makes no real attempt to strictly follow the chronological development of the SM. However, we try to indicate some of the important hurdles encountered along the way and how they were overcome. In addition, this approach gives us the opportunity to introduce in a natural fashion some of the basic concepts that underlie the Standard Model. We then briefly discuss some of the model's successes and limitations as well as some open questions that appear to require extensions beyond the model.

The modern view of the role of symmetry originated with Albert Einstein, whose general theory of relativity [3],[4] related gravitational interactions to the principle of general coordinate invariance. The use of a symmetry principle as the primary feature of nature that constrains and even dictates the allowable dynamical laws lies at the heart of the Standard Model. The Standard Model is designed to describe the strong, weak and electromagnetic interactions while treating gravity as a classical (flat) background. Thus it must respect the principles of special relativity. In addition, the dynamics needs to conform to the postulates of quantum theory. Since special relativity allows for particle pair production, any formulation must be able to account for arbitrary and varying numbers of particles. The only consistent way to achieve this is using a relativistic quantum field theory, with the textbook of Peskin and Schroeder [5] being one commonly used for reference. Here the particle modes arise as excitations of operator quantum fields $\zeta_a(x)$, labeled by the space-time point x^μ and an index a that distinguishes the various fields. (Appendix A.1 details the notations and convention employed.) The dynamics is encoded within the action functional

$$S[\zeta] = \int d^4x \mathcal{L}(x), \qquad (1.1)$$

where the local Lagrangian density $\mathcal{L}(x)$ is a function of the quantum fields and their space-time derivatives. This Lagrangian will also depend on various parameters. To fully define the theory, these parameters need to be fixed by certain normalization conditions. Using the Lagrangian, one extracts the various Feynman rules that can be employed in perturbative calculations. Some of the radiative corrections that arise from Feynman graphs containing closed loops can be included by defining scale-dependent (running) couplings to replace the couplings appearing in the Lagrangian. The form of the running coupling can be most easily secured using the renormalization group [5–9].

2 Wigner–Weyl Realization of Global Symmetries

Symmetries are represented by transformations of the fields that leave the action functional invariant. Included within this class are the space-time Poincaré transformations (space-time translations, spatial rotations and Lorentz boosts) of special relativity.

In addition to space-time symmetries, the action is constructed to be invariant under various transformations of the fields that depend on continuous parameters, $\{\omega_i\}$, which do not alter the space-time point. When the parameters themselves are independent of space-time, such symmetries are referred to as continuous global internal symmetries. In this case, a symmetry of the action is necessarily a symmetry of the Lagrangian density. Wigner's theorem [10] then dictates that these symmetries can be implemented on the Hilbert space of states by unitary operators, $U(\{\omega_i\})$, depending on parameters characterizing the transformation. Here $i = 1, \ldots, n$, with n being the number of group parameters. The invariance is then the statement that the Lagrangian density commutes with the unitary operator U:

$$U^{-1}(\{\omega_i\})\mathcal{L}(x)U(\{\omega_i\}) = \mathcal{L}(x). \qquad (2.1)$$

We shall primarily focus on continuous global transformations that form a Lie group. (See, for example, the text by H. Georgi [11].)

Although the Lagrangian $\mathcal{L}$ is constructed to be invariant under a certain set of transformations, the quantum fields will, in general, transform among themselves under these same transformations. We shall concentrate on quantum fields $\zeta_a(x)$ that transform irreducibly under the symmetry. The concept of irreducibility is simply that, under the transformation, a set of fields ζ_a will go into a linear combination of themselves. That is, one has

$$U^{-1}(\{\omega_i\})\zeta_a(x)U(\{\omega_i\}) = R_{ab}(\{\omega_i\})\zeta_b(x). \tag{2.2}$$

Here $R_{ab}(\{\omega_i\})$ are matrix elements of the representation matrices describing the given symmetry transformation.

Because we are dealing with continuous symmetries, it is possible and convenient to restrict attention to infinitesimal transformations. Finite transformations can then be obtained by simply compounding these infinitesimal transformations using the group property. For an infinitesimal transformation characterized by a set of dimensionless parameters $\{\delta\omega_i\}$, with $|\delta\omega_i| \ll 1$, one can write the operator $U(\{\delta\omega_i\})$ as

$$U(\{\delta a_i\}) = 1 + i\delta\omega_i G_i. \tag{2.3}$$

The operators G_i are known as the generators of the transformation. For each independent infinitesimal parameter $\delta\omega_i$ there is one generator associated with it. With no loss of generality, one can take these parameters to be real. The unitarity of the $U(\{\delta\omega_i\})$ then dictates that the generators G_i are Hermitian: $G_i = G_i^\dagger$. The group composition property associated with the symmetry transformations is encoded in the Lie algebra obeyed by the generators and characterized by a set of structure constants c_{ijk} so that

$$[G_i, G_j] = ic_{ijk}G_k. \tag{2.4}$$

The structure constants are obviously antisymmetric in i and j. Moreover, by suitably defining the generator basis, they can also be made totally antisymmetric in i, j and k. For infinitesimal transformations, the representation matrices R have a similar decomposition to Eq. (2.3) involving Hermitian matrices τ_i:

$$R_{ab}(\{\delta\omega_i\}) = \delta_{ab} + i\delta\omega_i(\tau_i)_{ab}. \tag{2.5}$$

These matrices provide a representation for the generators G_i and thus obey Eq. (2.4). Using Eqs. (2.4) and (2.5). It is easy to check that the transformation law for the fields ζ_a, given in Eq. (2.2), specifies the commutator of the generators G_i with the quantum fields being

$$[G_i, \zeta_a(x)] = -(\tau_i)_{ab}\zeta_b(x). \tag{2.6}$$

A general consequence of having a Lagrangian density $\mathcal{L}$ invariant under some global symmetry group is that associated with each of the independent symmetry transformations there exists a conserved operator (Noether) current [12] and a related time-independent operator charge. In fact, the operator charges are just the generators G_i, which for any given theory can be expressed in terms of the quantum fields ζ_a. The Noether current is given by

$$J_i^\mu(x) = \frac{\partial \mathcal{L}}{\partial \partial_\mu \zeta_a(x)} \frac{1}{i} (\tau_i)_{ab} \zeta_b(x), \tag{2.7}$$

which after application of the Euler–Lagrange field equation

$$\frac{\partial \mathcal{L}}{\partial \zeta_a(x)} - \partial_\mu \frac{\partial \mathcal{L}}{\partial \partial_\mu \zeta_a(x)} = 0 \tag{2.8}$$

is found to be conserved:

$$\partial_\mu J_i^\mu(x) = 0. \tag{2.9}$$

The associated time-independent charge is

$$G_i = \int d^3x\, J_i^0(x). \tag{2.10}$$

Provided that the vacuum state of the theory is left invariant by all of the symmetry transformations so that

$$U(\{\omega_i\})|0\rangle = |0\rangle, \tag{2.11}$$

or equivalently

$$G_i|0\rangle = 0, \tag{2.12}$$

there are mass degenerate multiplets of states in the physical spectrum. When both the Lagrangian and the vacuum state are invariant under the global symmetry transformation, the symmetry is said to be realized à la Wigner–Weyl [13, 14].

To establish that in the case of a Wigner–Weyl realized global symmetry there are mass degenerate multiplets of states in the theory, we need to recall some properties of single-particle states. For each of the quantum fields $\zeta_a(x)$, one can effectively associate a single-particle state $|p; \alpha\rangle$, with $p^\mu = (\sqrt{\vec{p}^2 + m_\alpha^2}, \vec{p})$, by the Lehmann, Symanzik, Zimmermann (LSZ) prescription [15]. For a scalar field ζ_a, for example, one writes

$$|p; a\rangle = \lim_{x^0 \to \pm\infty} \frac{-i}{\sqrt{Z}} \int d^3x\, e^{ipx} \overset{\leftrightarrow}{\partial}_0 \chi_a(x)|0\rangle, \tag{2.13}$$

where $|0\rangle$ is the vacuum state and $\langle 0|\chi_a(0)|p; \alpha\rangle = \sqrt{Z}$. (Since for single-particle states, there is no distinction between in and out states, one can take the limit either as x^0 goes to $+\infty$ or $-\infty$.) The notation px here is a shorthand for $p^\mu x_\mu$, while $A\overset{\leftrightarrow}{\partial}_0 B \equiv A\partial_0 B - (\partial_0 A)B$. Acting with $U^{-1}(\{\omega_i\})$ on the preceding expression gives

$$U^{-1}(\{\omega_i\})|p; a\rangle = \lim_{x^0 \to \pm\infty} \frac{-i}{\sqrt{Z}} \int d^3x e^{ipx} \overset{\leftrightarrow}{\partial}_0 U^{-1}(\{\omega_i\})\chi_a(x)|0\rangle. \tag{2.14}$$

If one assumes that the vacuum state is invariant under the symmetry transformation, so that

$$U(\{\omega_i\})|0\rangle = |0\rangle, \tag{2.15}$$

then the unitary operator $U(\{\omega_i\})$ can be inserted with impunity between the vacuum state and $\chi_a(x)$. In this case, one can use Eq. (2.2) to immediately deduce that the single-particle states $|p; a\rangle$ transform as

$$U^{-1}(\{\omega_i\})|p; a\rangle = R_{ab}(\{\omega_i\})|p; \beta\rangle. \tag{2.16}$$

That is, these states transform irreducibly among themselves when acted upon by the unitary operator U^{-1}.

The Hamiltonian of the theory, H, because of the invariance of $\mathcal{L}$ under the symmetry transformations, commutes with $U^{-1}(\{\omega_i\})$:

$$[H, U^{-1}(\{\omega_i\})] = 0. \tag{2.17}$$

By definition, the action of H on a single-particle state at rest gives the mass of that state

$$H|p; a\rangle_{\text{rest}} = m_a|p; a\rangle_{\text{rest}}. \tag{2.18}$$

Using the preceding equations, it follows that

$$0 = [H, U^{-1}(\{\omega_i\})] |p; a\rangle_{\text{rest}} = R_{ab}(\{\omega_i\})(m_a - m_b)|p; b\rangle_{\text{rest}}. \tag{2.19}$$

Because the preceding must hold for all transformations $R_{ab}(\{\omega_i\})$, it follows that

$$m_a = m_b. \tag{2.20}$$

Thus, provided that the vacuum state is invariant under the symmetry, then each irreducible representation corresponds to a multiplet of single-particle states degenerate in mass.

3 Local $U(1)$ Invariance and QED

An explicit example of a continuous global invariance is afforded by the free field Dirac action

$$S_{Dirac}[\psi, \overline{\psi}] = \int d^4x \mathcal{L}_{Dirac}(x), \tag{3.1}$$

with associated Lagrangian

$$\mathcal{L}_{Dirac}(x) = -\overline{\psi}(x) \left(\gamma^\mu \frac{1}{i} \partial_\mu + m \right) \psi(x), \tag{3.2}$$

whose Euler–Lagrange equations are simply the free Dirac equation

$$\left(\gamma^\mu \frac{1}{i}\partial_\mu + m\right)\psi(x) = 0 \tag{3.3}$$

and its Hermitian conjugate. Said Lagrangian is invariant under the global, continuous phase transformation

$$\psi(x) \rightarrow e^{iq_\psi \omega}\psi(x); \quad \overline{\psi}(x) \rightarrow e^{-iq_\psi \omega}\overline{\psi}(x), \tag{3.4}$$

which corresponds to a $U(1)$ Abelian group. (In this case, the structure constants vanish: $c_{ijk} = 0$.) Here ω is a real space-time–independent parameter and q_ψ is a real number that for electrons is chosen by convention (which can be traced to Benjamin Franklin) as $q_e = -1$. For infinitesimal parameter $\delta\omega$, $\psi(x) \rightarrow \psi(x) + \delta\psi(x)$ with $\delta\psi(x) = iq_\psi \delta\omega\psi(x)$. Similarly $\delta\overline{\psi}(x) = -iq_\psi \delta\omega\overline{\psi}(x)$. This invariance and the application of the field equations leads via Noether's theorem to the conserved electromagnetic current operator

$$J_{EM}^\mu(x) = q_\psi \overline{\psi}(x)\gamma^\mu \psi(x) = -\overline{\psi}(x)\gamma^\mu \psi(x), \tag{3.5}$$

with $\partial_\mu J_{EM}^\mu(x) = 0$ and the time-independent electric charge operator

$$Q_{EM} = \int d^3r J_{EM}^0(x). \tag{3.6}$$

Since the vacuum carries zero electric charge so that $Q_{EM}|0\rangle = 0$, it follows that this symmetry is realized à la Wigner–Weyl and results in a mass degeneracy. In this case, the degeneracy is between the electron and positron masses.

Now suppose one allows the transformation parameter to vary from point to point in space-time so that $\omega = \omega(x)$. Under this local or gauged $U(1)$ phase transformation,

$$\psi(x) \rightarrow e^{iq_\psi \omega(x)}\psi(x); \quad \overline{\psi}(x) \rightarrow e^{-iq_e\omega(x)}\overline{\psi}(x), \tag{3.7}$$

the free Dirac Lagrangian is no longer invariant:

$$\mathcal{L} \rightarrow \mathcal{L} - J_{EM}^\mu \partial_\mu \omega. \tag{3.8}$$

In order to restore the invariance, one simply replaces the derivative by a covariant derivative

$$\partial_\mu \rightarrow D_\mu = \partial_\mu - ieA_\mu(x), \tag{3.9}$$

where the four-vector potential $A_\mu(x)$ transforms as

$$eA_\mu(x) \rightarrow eA_\mu(x) + \partial_\mu \omega(x). \tag{3.10}$$

Here we have introduced the QED coupling e. Demanding local or gauge invariance then results in the action of Quantum Electrodynamics (QED).

(For selected fundamental papers in the development of QED, see, for example, reference [16].)

$$S_{QED}[\psi, \overline{\psi}, A^\mu] = \int d^4x \mathcal{L}_{QED}(x), \tag{3.11}$$

where

$$\mathcal{L}_{QED}(x) = \mathcal{L}_{Dirac}(x) + eJ_{EM}^\mu(x)A_\mu(x) + \mathcal{L}_{Maxwell}(x). \tag{3.12}$$

Here the Maxwell Lagrangian

$$\mathcal{L}_{Maxwell}(x) = -\frac{1}{4}F^{\mu\nu}(x)F_{\mu\nu}(x), \tag{3.13}$$

with gauge-invariant field strength

$$F_{\mu\nu}(x) = \partial_\mu A_\nu(x) - \partial_\nu A_\mu(x), \tag{3.14}$$

provides the kinetic term for the photon field. Thus we extract the main lesson that the interaction term between the electron field and the photon field arises as a consequence of the imposition of the local or gauge invariance. For QED, the interaction is the coupling of the conserved electromagnetic vector current to the vector photon field: $eJ_{EM}^\mu(x)A_\mu(x)$. Any field carrying an electric charge contributes to the current and directly couples to the photon. Note that since photons themselves carry zero electric charge, they do not have any direct self-interactions. By the late 1940s, R. P. Feynman, J. Schwinger, S. I. Tomonaga and F. Dyson established [16] that QED could be consistently interpreted order by order in perturbation theory once the coupling and mass of the electron were fixed as inputs.

Theoretical calculations in QED can be performed perturbatively in powers of the fine structure constant $\alpha = \frac{e^2}{4\pi}$, whose numerical value at the electron mass scale is experimentally extracted as [17] $\alpha^{-1} = 137.035999206(81)$. At present, these calculations have been carried out with exquisite precision and exhibit remarkable agreement with electrodynamic measurements, which have been performed with incredible accuracy. For example, consider the single-electron matrix element of the electromagnetic current

$$\langle e^-(p_1)|J_{EM}^\mu(0)|e^-(p_2)\rangle = \overline{U}_e(p_1)\left(\gamma^\mu F_1(q^2) + \sigma^{\mu\nu}q_\nu F_2(q^2)\right)U_e(p_2), \tag{3.15}$$

which encodes the electron electromagnetic properties. This decomposition is a general consequence of Lorentz invariance and the fact that the electromagnetic current J_{EM}^μ is conserved. Here $|e^-(p_1)\rangle$ and $|e^-(p_2)\rangle$ are single-electron states with energy-momentum $p_1^\mu = \left(\sqrt{\vec{p}_1^2 + m^2}, \vec{p}_1\right)$ and $p_2^\mu = \left(\sqrt{\vec{p}_2^2 + m^2}, \vec{p}_2\right)$

respectively and $q^\mu = p_2^\mu - p_1^\mu$ is the momentum transfer. Note that the form of this decomposition holds if the electron is replaced by any single fermion state. Moreover, it is applicable in any theory where there is a conserved electromagnetic current. This is true if the electromagnetic current is associated with a gauged symmetry as in QED or is an external current as, for example, in the theory of strong interactions, QCD. All that will change are the form factors $F_1(q^2)$ and $F_2(q^2)$. In some cases, such as QED, these form factors can be computed perturbatively. In other cases, like the theory of strong interactions, the computation of the form factors, in general, requires a nonperturbative analysis.

The physical interpretation of the form factors at zero momentum transfer is secured by further coupling the electron to an external static electromagnetic field. From the coupling to the external electric field, one deduces that $F_1(0)$ is the charge on the electron. Since Q_{EM} generates the electric charge symmetry, it follows that

$$[Q_{EM}, \overline{\psi}(x)] = \overline{\psi}(x), \tag{3.16}$$

which acting on the vacuum gives

$$Q_{EM}\overline{\psi}(x)|0\rangle - \overline{\psi}(x)Q_{EM}|0\rangle = \overline{\psi}(x)|0\rangle. \tag{3.17}$$

Using $Q_{EM}|0\rangle = 0$, it follows that

$$Q_{EM}|e^-(p_1)\rangle = |e^-(p_1)\rangle. \tag{3.18}$$

Taking the matrix element with $\langle e^-(p_1)|$ and using the preceding decomposition, Eq. (3.15), in conjunction with the fact that the electromagnetic current is conserved so the time-independent electric charge operator is $Q_{EM} = \int d^3r \, J_{EM}^0(x)$, then yields

$$F_1(0) = q_e = -1, \tag{3.19}$$

which is the experimentally observed value. Note that if the electron states are replaced by muon states, one again finds $F_1(0) = -1$, while if electron states are replaced by proton states, $F_1(0) = 1$, which gives the electric charge of the proton. Note that the value of $F_1(0)$ for the electron is unchanged by the higher-order perturbative QED interactions or in the proton case by electromagnetic or strong interactions. That is, the charge form factor at zero momentum transfer does not get renormalized. This is a consequence of electromagnetic current conservation. Moreover, the fact that $F_1(0)$ has the same magnitude for electrons, muons and protons reflects a universality of the QED interaction.

On the other hand, the coupling to an external magnetic field dictates that the magnetic moment operator for the electron is

$$\vec{\mu}_e = -\mu_B g_e \frac{1}{2}\vec{\sigma}, \tag{3.20}$$

where the $\vec{\sigma}$ are the three 2×2 Pauli matrices. Here

$$\mu_B = \frac{e}{2m} \tag{3.21}$$

is the Bohr magneton and the electron g_e factor is

$$g_e = 2\left(1 + F_2(0)\right) \tag{3.22}$$

so that

$$F_2(0) = \frac{1}{2}(g_e - 2) \equiv a_e. \tag{3.23}$$

Free Dirac theory (coupled to the external EM field) gives $F_2(0)|_{Dirac} = a_e|_{Dirac} = 0$, thus accounting for the dominant contribution $g_e|_{Dirac} = 2$, which was one of its major successes. However, radiative (loop) corrections due to QED modify this result, producing a nontrivial anomalous magnetic moment a_e. The 1-loop $O(\alpha)$ theoretical calculation giving $a_e^{1\ loop} = \frac{\alpha}{2\pi}$ was first performed by J. Schwinger [18] in 1948. Subsequently the calculation has been extended to include terms analytically [19] of $O(\alpha^3)$ and numerically through $O(\alpha^5)$ yielding [20]

$$a_e^{th} = 0.001\ 159\ 652\ 181\ 61\ (23). \tag{3.24}$$

(Note that this value includes the contributions from the weak and strong interactions in addition to the pure QED result.) Experimentally, the anomalous magnetic moment of the electron has been measured [21] as

$$a_e^{exp} = 0.001\ 159\ 652\ 180\ 73\ (28), \tag{3.25}$$

so there is agreement to 10 significant figures. Both the experimental measurement and theoretical calculation of this anomalous magnetic moment of the electron are truly monumental achievements. Their agreement certainly provides credence to the validity of QED.

4 Global Symmetry Structure of the Strong Interactions

While no viable theory of the strong nuclear force holding the nucleus together yet existed, nonetheless, shortly after the 1932 J. Chadwick discovery [22] of the neutron with a mass nearly degenerate with that of the proton, W.

Heisenberg [23] proposed that this near-mass degeneracy could be accounted for if the interactions exhibit an $SU(2)_V$ global vector symmetry realized à la Wigner–Weyl with the proton and neutron transforming as a doublet. The charge operators, T_i $i = 1, 2, 3$, of this non-Abelian strong isotopic symmetry satisfy the $SU(2)$ Lie algebra

$$[T_i, T_j] = i\epsilon_{ijk}T_k, \tag{4.1}$$

where ϵ_{ijk} is the Levi–Civita tensor satisfying $\epsilon_{123} = 1$. The associated conserved vector currents are $SU(2)_V$ generalizations of the conserved electromagnetic current except now the currents carry a net electric charge. The small explicit breaking of the isospin symmetry producing the relatively small neutron-proton mass difference is today recognized as resulting from a combination of the up-down quark mass differences and the perturbative electromagnetic interactions, which affects differently the charged and neutral components of the multiplet. The origin of each of these effects lies outside the strong interaction. Thus the strong interaction itself respects the global $SU(2)_V$ isospin symmetry. After the introduction of a new conserved quantum number dubbed strangeness following the discovery of the neutral and charged spin zero kaons and the baryonic Λ, this global classification symmetry was extended to an $SU(3)_V$ symmetry independently by M. Gell-Mann [24] and Y. Ne'eman [25] in 1961. (There is far larger mass splitting among the components of mesonic and baryonic $SU(3)_V$ multiplets relative to that within the $SU(2)_V$ multiplets. Today, this is recognized as due to the s quark mass being larger than the u and d masses.)

The charged and neutral pions are the lowest-mass hadronic states. They transform as a triplet under the global $SU(2)_V$ isospin symmetry, which explains the near-degeneracy of the $\pi^\pm$ and π^0 masses. It was established that many of the pion properties could be accounted for by assuming there existed partially conserved axial vector currents (PCAC) [26] with the small explicit symmetry breaking being proportional to the square of the pion mass. The origin of this small explicit breaking mass is due to the up and down quark masses and thus again lies outside the strong interaction. Consequently, the strong interactions have a conserved global $SU(2)_A$ axial symmetry and associated conserved axial vector currents. However, since no near-mass-degenerate opposite parity states exist, it follows that this symmetry is not realized à la Wigner–Weyl. The understanding of the way this symmetry is realized proved pivotal to the understanding of the weak interaction and hence the construction of the Standard Model.

5 Fermi Theory of the Weak Interactions

Attempting to model the short-ranged ($\sim 10^{-18}$ m) weak nuclear force responsible for nuclear beta decay data, E. Fermi [27–29] introduced in 1933 a local current-current coupling. The currents were taken as vector currents in analogy with the current appearing in electrodynamics, but they were electric-charge-carrying currents. The Fermi model conserved parity. However in 1956, C. N. Yang and T. D. Lee raised [30] the possibility that parity might be violated by the weak interactions. The experiment they proposed to test this hypothesis was successfully performed by C.-S. Wu and her collaborators [31] in 1957 with the conclusion that parity was indeed violated by the weak interactions. To account for this parity violation, R. Feynman and M. Gell-Mann [32] and independently R. Marshak and E. C. G. Sudarshan [33] modified the Fermi model so that charged currents were of a vector-axial vector (V-A) form

$$J^\mu_{cc} = V^\mu_{cc} - A^\mu_{cc} \qquad (5.1)$$

and the modified Fermi theory interaction Lagrangian reads

$$\mathcal{L}_{Fermi} = G_F J^\mu_{cc} J_{cc\,\mu}. \qquad (5.2)$$

Here $G_F \sim 10^{-5}$ GeV^{-2} is the Fermi constant, which sets the scale of the weak interactions, and the currents were composed of a sum of hadronic and leptonic pieces. Moreover, they and independently S. Gershtein and Ya. B. Zeldovich [34] proposed that the hadronic piece of the vector current is the conserved vector current of the global isotopic spin $SU(2)_V$ symmetry of strong interactions introduced by Heisenberg. This is referred to as the conserved vector current (CVC) hypotheis. A bit later, M. Gell-Mann further proposed [26] that the hadronic piece of the axial vector current is the conserved axial vector current of the global $SU(2)_A$ symmetry of the strong interaction. The current-current interaction could account reasonably well at the time for the nuclear beta decay data using lowest-order perturbation theory since the amplitudes varied as $G_F E^2$ and the energies were less than 1 GeV. However, the model was internally inconsistent once higher-order (loop) corrections were included.

As early as 1948, attempting to more closely mimic the successful QED interaction, O. Klein [35] suggested replacing the local current-current Fermi model with a model that coupled the charged weak interaction currents to charged intermediate vector boson fields, $W^\pm_\mu$, as

$$\mathcal{L}_{cc} = g_W(J^\mu_{cc} W^+_\mu + J^{\dagger\mu}_{cc} W^-_\mu). \qquad (5.3)$$

The vector particle excitations, $W^{\pm}$, of the vector fields need to be far more massive than the momentum exchanged to account for the short-range nature of the weak interaction processes probed at that time. To reproduce the results of the Fermi theory, one simply approximates the massive vector (Proca) propagator by its ultralocal form ($M_W^2 \gg q^2$),

$$\Delta_{\mu\nu}(q) = \frac{\eta_{\mu\nu} + \frac{q_\mu q_\nu}{M_w^2}}{q^2 + M_W^2 - i\epsilon} \rightarrow \frac{\eta_{\mu\nu}}{M_W^2}, \tag{5.4}$$

while identifying $G_F \sim \frac{g_W^2}{M_W^2}$. Here, $\eta_{\mu\nu}$ is the Minkowski space metric tensor. Taking $g_W \sim 10^{-1}$, which is the strength of the electromagnetic coupling, gives $M_W \sim 100$ GeV. Unfortunately, simply mimicking the form of the QED interaction is not enough to cure the model of its sicknesses. The root of the problem lies in the origin of the vector mass. Simply adding a direct $W^{\pm}$ mass term to the model Lagrangian yields a Proca vector propagator, which does not fall off at large energies as does the massless photon propagator in QED. Thus, as in the local current-current Fermi theory, unitarity is violated at energies of the order of 300 GeV.

6 Nambu–Goldstone Realization of $SU(2)_A$

A significant advance in the understanding of the realization of the global $SU(2)_A$ symmetry of the strong interactions was achieved by Y. Nambu [38] who was inspired by the J. Bardeen, L. Cooper, R. Schrieffer (BCS) theory [36],[37] of superconductivity, which accounts for the attractive force leading to bound Cooper pairs of electrons in a spin singlet state as arising from their interaction with phonons in the lattice. Due to the condensate of Cooper pairs, the BCS ground state does not exhibit the full symmetry structure of the BCS Hamiltonian. In 1960, Nambu suggested that the vacuum state of the strong interactions, in analogy to the superconducting ground state, might not respect the full symmetries of the theory, and the elementary particles might acquire mass in a manner analogous to the energy gap of the quasiparticle excitations of BCS theory. To study this conjecture, Nambu and G. Jona-Lasinio [39] introduced a Lorentz invariant model, the NJL model, in which the Lagrangian is invariant under independent vector and axial vector global $U(1)$ phase transformations. This is tantamount to having independent phase transformations for the left-handed and right-handed components of the fermion field:

$$f_L(x) = \frac{1}{2}(1 - \gamma_5)f(x),$$

$$f_R(x) = \frac{1}{2}(1 + \gamma_5)f(x). \tag{6.1}$$

Invariance under the axial transformation forbids a direct fermion mass term. This NJL model, whose interaction term was a chirally invariant four-fermion coupling, was self-consistently analyzed in mean field approximation. A nontrivial vacuum fermion-antifermion condensate emerged, which broke the axial symmetry and resulted in a nonzero fermion mass. Thus while the model Lagrangian is invariant under the axial symmetry, the vacuum state is not. The symmetry is said to be spontaneously broken. (The nomenclature was introduced by S. Glashow and M. Baker [41]. Unfortunately it is not an accurate description of what is going on, as the symmetry is not really broken. A better name is hidden or secret symmetry as coined by S. Coleman [42]. Nonetheless, the spontaneous symmetry breaking moniker stuck.) In addition, the model predicted the appearance of a massless composite pseudoscalar particle whose interactions were derivatively coupled.

A very useful phenomenological description of superconductivity predating BCS is the Ginzburg–Landau Model [43]. The model introduces an electrically charged scalar order parameter field $\phi(\vec{r})$ representing the wavefunction of the condensate of Cooper pairs whose Hamiltonian,

$$H_{GL} = \phi^*(\vec{r})\frac{1}{2m}\left(\frac{1}{i}\nabla - 2e\vec{A}(\vec{r})\right) + V(\phi^*\phi), \tag{6.2}$$

with potential

$$V(\phi^*\phi) = \beta\phi^*\phi + \frac{1}{2}\gamma(\phi^*\phi)^2 \tag{6.3}$$

respects a U(1) gauge invariance. The coefficients, β and γ, are temperature dependent with β changing sign at the critical temperature, T_c, going negative for $T < T_c$. Thus for these temperatures the potential minimum is at $\phi^*\phi = -\frac{\beta}{\gamma}$. The magnitude of the order parameter is fixed, but the phase is arbitrary corresponding to a family of degenerate ground states. The selection of a particular ground state corresponds to a spontaneous breaking of the U(1) symmetry of the Hamiltonian by the ground state. Thus this nonrelativistic model exhibits a spontaneous symmetry breaking and is a precursor to the Lorentz–invariant U(1) Goldstone and Higgs models we shall soon address.

An earlier example of a spontaneously broken global symmetry is afforded by the W. Heisenberg description [44] of ferromagnetism using a Hamiltonian,

$$H_{\text{Heisenberg}} = -\frac{1}{2}\sum_{x,y} J_{x-y}\vec{S}_x \cdot \vec{S}_y, \tag{6.4}$$

where $\vec{S}_x$ denotes the spin at site x and J is the strength of the interaction, which depends on the distance between the sites x and y and is assumed to fall rapidly as the distance becomes large. The model is invariant under a simultaneous rotation of all the spins. For positive J, the lowest energy state is one with all spins aligned; the particular direction is arbitrary. Thus there is a degeneracy of ground states. From one such ground state, one can make a rotation of all the spins and get another ground state degenerate in energy. Once a particular ground state is chosen, the rotational symmetry of the Hamiltonian is spontaneously broken. The model exhibits an excitation, the spin wave, whose energy vanishes in long wavelength limit.

While the Nambu–Jona–Lasinio model yielded a nonzero fermion mass, it did not prove to be the correct theory of strong interactions. In fact, the model was actually inconsistent when higher-order radiative corrections were taken into account. (It is nonrenormalizable.) An alternative model exhibiting a global $SU(2)_V \times SU(2)_A$ symmetry where the axial symmetry is spontaneously broken, resulting in a nonzero fermion mass was suggested by J. Schwinger [40]. He studied aspects of what later would be called the Gell-Mann–Levy σ model [45]. The model degrees of freedom consisted of an $SU(2)_V$ nucleon doublet N (proton and neutron), the three pions, $\vec{\pi}$, forming an $SU(2)_V$ triplet and a hypothetical $SU(2)_V$ singlet field σ. The $SU(2)_V \times SU(2)_A$ symmetry is realized as

$$N(x) \rightarrow \left(1 + i\delta\vec{\omega}_V \frac{\vec{\sigma}}{2} + i\delta\vec{\omega}_A \cdot \frac{\vec{\sigma}}{2}\gamma_5\right) N(x),$$

$$\sigma(x) \rightarrow \sigma(x) + \delta\vec{\omega}_A \cdot \vec{\pi}(x),$$

$$\vec{\pi}(x) \rightarrow \vec{\pi}(x) - \delta\vec{\omega}_V \times \vec{\pi}(x) - \delta\vec{\omega}_A\sigma(x), \tag{6.5}$$

and forbids a fermion mass term. Here the $\vec{\sigma}$ are the three 2×2 Pauli matrices. The chirally invariant interaction is of the Yukawa form:

$$\mathcal{L}_{\text{Yukawa}} = g_{\text{Yukawa}}\overline{N}(\sigma + i\vec{\sigma} \cdot \vec{\pi}\gamma_5)N. \tag{6.6}$$

If the σ field develops a nontrivial vacuum expectation value, $\langle\sigma\rangle \neq 0$, the fermion mass $m_N = g_{\text{Yukawa}}\langle\sigma\rangle$ is generated. This Schwinger mechanism of fermion mass generation is basically how fermion masses are produced in the standard electroweak theory.

At roughly the same time as Nambu's work, J. Goldstone [46] studied the simplest Lorentz-invariant model exhibiting a spontaneous global continuous symmetry breaking. It consisted of a single complex scalar field $\phi(x)$ whose dynamics is invariant under a global $U(1)$ phase transformation $\phi(x) \rightarrow e^{i\omega}\phi(x)$. The invariant Lagrangian is

$$\mathcal{L}_{\text{Goldstone}} = -\partial_\mu\phi^\dagger(x)\partial^\mu\phi(x) - V(\phi) \tag{6.7}$$

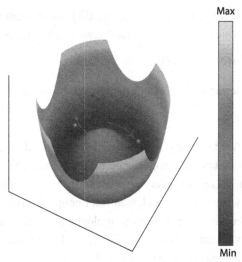

Max

Min

Figure 6.1 Sombrero potential: massless mode rolls along bottom of well; massive mode sloshes up and down wall. The colorbar represents the value of the potential value ranging from its minimum (dark blue) to its maximum (yellow).

with

$$V(\phi) = \mu^2 \phi^\dagger(x)\phi(x) + \frac{1}{2}\lambda\left(\phi^\dagger(x)\phi(x)\right)^2. \tag{6.8}$$

So long as $\mu^2 > 0$, the potential is minimized at $\phi = 0$ and the spectrum consists of a charged scalar and its antiparticle degenerate in mass, namely a Wigner–Weyl realization. But for $\mu^2 < 0$, the potential takes the form of a sombrero as displayed in Figure 6.1. In this case, the potential is minimized for $\phi^\dagger\phi = \frac{v^2}{2}$, where $v^2 = -\frac{\mu^2}{\lambda}$. Hence the lowest energy states are such that ϕ has magnitude $\frac{v}{\sqrt{2}}$ but arbitrary phase and the vacuum state is infinitely degenerate. Once the phase is fixed (the particular value is physically unobservable), the global symmetry is spontaneously broken. Choosing the phase to vanish so $\langle\phi\rangle = v$ is real and expanding about that point as

$$\phi(x) = \frac{1}{\sqrt{2}}(v + \phi_1(x) + i\phi_2(x)) \tag{6.9}$$

with ϕ_1, ϕ_2 Hermitian fields, the model describes two spin zero particles, one with mass $m_{\phi_1} = \sqrt{\lambda}v$, while the other is massless, $m_{\phi_2} = 0$. Once again, there is a massless spin zero particle associated with a spontaneously broken continuous global symmetry. The massless particle is referred to as a Nambu–Goldstone boson (NGB) and the spontaneously broken symmetry constitutes a Nambu–Goldstone realization of the continuous global symmetry. In

Nambu–Goldstone realizations, the action is still invariant under the symmetry transformations, which continue to be represented by unitary operators so that

$$U^{-1}(\{\omega_i\})\mathcal{L}(x)U(\{\omega_i\}) = \mathcal{L}(x). \tag{6.10}$$

Now, however, the vacuum state does not respect the symmetry, so

$$U(\{\omega_i\})|0\rangle \neq |0\rangle. \tag{6.11}$$

J. Goldstone, A. Salam and S. Weinberg [47] provided a general proof of a theorem which dictates the appearance of a massless, spin zero, Nambu–Goldstone boson (NGB) for every spontaneously broken continuous global internal symmetry in any model with manifest Lorentz invariance. The proof is independent of the particular dynamics employed to spontaneously break the symmetry. This constituted an apparent death knoll for all such models, as no such massless spin 0 particles were observed.

7 Yang–Mills Theory

C. N. Yang and R. Mills [48] were the first[1] to construct a mathematically consistent four space-time dimensional theory containing spin-one particles which carry nontrivial gauge group quantum numbers. To do so requires a non-Abelian gauge theory that generalizes the Abelian U(1) gauge theory of QED to non-Abelian gauge groups. (In their original paper, they focused on an $SU(2)$ group. The more general case soon followed.) This involves promoting the local phase invariance of the Abelian model to a matrix-valued local phase invariance. Consider a non-Abelian gauge group with Lie algebra of dimension n, which is also the number of group generators G_i, group parameters $\omega_i(x)$ and vector gauge fields A_i^μ where $i = 1, 2, \ldots, n$. It proves convenient to define the matrix-valued field $A^\mu = A_i^\mu \tau_i$, where τ_i are matrix representations of the group generators G_i that satisfy the non-Abelian algebra

$$[G_i, G_j] = ic_{ijk}G_k \tag{7.1}$$

with group structure constants c_{ijk}. It follows that the A^μ transformation takes the form

$$A_\mu(x) \rightarrow R(\omega(x))A_\mu(x)R^{-1}(\omega(x)) - \frac{i}{g}R(\omega(x))\partial_\mu R^{-1}(\omega(x)), \tag{7.2}$$

where $R(\omega(x)) = e^{i\omega_i(x)\tau_i}$ is a matrix representation of the non-Abelian group element.

---

[1] In 1954, R. Shaw, a graduate student of A. Salam at the University of Cambridge, performed similar work to Yang–Mills. However, this work was unpublished and appears only in Shaw's Ph.D. thesis.

Yang and Mills constructed the invariant Lagrangian

$$\mathcal{L}_{YM}(x) = -\frac{1}{4}F_{\mu\nu\,i}(x)F_i^{\mu\nu}(x), \tag{7.3}$$

where

$$F_i^{\mu\nu}(x) = \partial^\mu A_i^\nu(x) - \partial^\nu A_i^\mu(x) + gc_{ijk}A_j^\mu(x)A_k^\nu(x) \tag{7.4}$$

is the Yang–Mills field strength, which transforms as the adjoint representation. Contrary to the QED photon field, the non-Abelian vectors are self-coupled and the Yang–Mills Lagrangian contains three- and four-point vector interactions whose couplings are related in a definite way.

Note, however, that the mass term $A_i^\mu(x)A_{\mu\,i}(x)$ is not invariant under the above gauge transformation and thus is forbidden. Consequently, it would appear that to maintain the non-Abelian gauge symmetry, the non-Abelian vectors are necessarily massless. Thus the theory was initially considered a failure since it seemed incapable of explaining the short-ranged weak interactions nor did it seem relevant for the strong interactions. As it turned out, Yang–Mills theories provide the proper framework to describe both the strong and electroweak interactions.

8 Quantum Chromodynamics

Quantum Chromodynamics (QCD) as the theory of strong interactions was introduced in the early 1970s by M. Gell-Mann, H. Fritzsch, W. Bardeen, and H. Leutwyler [49], [50–52]. It is an unbroken $SU(3)_c$ gauge theory where the local symmetry involves transformations in the space of a new quantum number dubbed color. The Yang–Mills fields mediating the interaction are the massless self-interacting vector gluon fields, $A_i^\mu(x)$, $i = 1, 2, \ldots 8$, which transform as in Eq. (7.2) and also interact with the fermionic spin 1/2 quark fields which transform as the fundamental (triplet) representation. The QCD Lagrangian is

$$\mathcal{L}_{QCD} = \sum_\alpha \mathcal{L}(q_a^\alpha, D_\mu q_a^\alpha) - \frac{1}{4}G_i^{\mu\nu}G_{\mu\nu i}, \tag{8.1}$$

where

$$\mathcal{L}(q_a^\alpha, D_\mu q_a^\alpha) = -\bar{q}_a^\alpha \gamma^\mu \frac{1}{i} D_\mu q_a^\alpha - m_\alpha \bar{q}_a^\alpha q_a^\alpha, \tag{8.2}$$

with covariant derivative

$$D_\mu q_a^\alpha = \left(\delta_{ab}\partial_\mu - ig_3\left(\frac{\lambda_i}{2}\right)_{ab} A_{\mu i}\right) q_b^\alpha. \tag{8.3}$$

The gluon field strength,

$$G_i^{\mu\nu} = \partial^\mu A_i^\nu - \partial^\nu A_i^\mu + g_3 f_{ijk}A_j^\mu A_k^\nu, \tag{8.4}$$

transforms as the adjoint (octet) representation under $SU(3)_c$. In the preceding equations, $\alpha = \{u, d, c, s, t, b\}$ is a quark flavor label, while $a = 1, 2, 3$, and $i = 1, 2, .., 8$, are color labels. The λ_i are the Gell-Mann matrices, which form the 3×3 fundamental representation for $SU(3)_c$, while the f_{ijk} are the $SU(3)_c$ group structure constants and g_3 is the $SU(3)_c$ gauge coupling.

While the role of a color quantum as a necessary property for the spin 1/2 quarks was emphasized very early [53–56] in order for the quark model for baryons to be consistent with the spin-statistics theorem, the dynamical importance of having a non-Abelian local symmetry describing the interactions of quarks did not really become manifest until the discovery of asymptotic freedom by D. Gross and F. Wilczek [57] and independently by H. D. Politzer [58] in 1973 who demonstrated that due to the non-Abelian QCD vector gluon self-couplings, the interaction strength (logarithmically) decreases at shorter and shorter distances or equivalently at higher and higher energies. Note that this running of the QCD coupling is opposite to the case of the Abelian QED where the interaction strength decreases at longer distances due to electric charge screening.

Explicitly, D. Gross, F. Wilczek and H. D. Politzer computed the 1-loop running of the gauge coupling in a Yang–Mills theory with gauge group G and n_R fermions carrying gauge representation R and showed that asymptotic freedom emerges provided

$$\sum_R n_R T(R) < \frac{11}{4} C_2(G). \tag{8.5}$$

Here $C_2(G)$ is the value of the quadratic Casimir operator for the adjoint representation of the gauge group G and $T(R)$ is given by

$$r T(R) = d(R) C_2(R), \tag{8.6}$$

with $C_2(R)$ is the value of the quadratic Casimir operator for the representation R, $d(R)$ the dimension of the representation R and r the dimension (number of generators) of the group G. For an $SU(N)$ group, $C_2(SU(N)) = N$ and $d = N^2 - 1$, while for fermions transforming as the fundamental representation of $SU(N)$ with dimension $d(N) = N$, the Casimir operator is $C_2(N) = \frac{N^2-1}{2N}$ so that $T(N) = \frac{1}{2}$. In this case, the condition for asymptotic freedom, Eq. (8.5) is

$$\frac{n_g}{2} < \frac{11}{4} N, \tag{8.7}$$

which dictates that QCD with gauge group $SU(3)$ will be asymptotically free, provided that there are less than $\frac{33}{2}$ quark flavors transforming as the fundamental representation. This condition is clearly satisfied by the currently observed

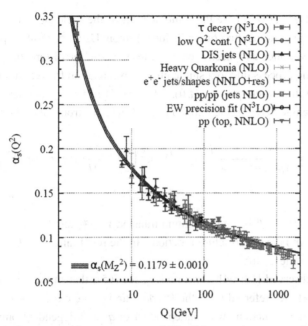

Figure 8.1 Summary of measurements [62] of the running α_s as a function of the energy scale Q. The agreement between the data and the QCD expectations is apparent. The order of QCD perturbation theory used in the extraction of α_s is indicated in brackets (NLO: next-to-leading order; NNLO: next-to-next-to-leading order; NNLO+res.: NNLO matched to a re-summed calculation; N^3LO: next-to-NNLO).

six such flavors. On the other hand, $SU(3)$ Yang–Mills theory with more than two octets of fermions is not asymptotically free.

Note that an analogous running of the gauge coupling had previously been observed in quantum electrodynamics coupled to a charged vector field, by V. S. Vanyashin and M. V. Terent'ev [59] in 1965 and in pure Yang–Mills theory by I. Khriplovich [60] in 1970 and G. 't Hooft [61] in 1972, but its physical significance was not fully realized until the work of Gross, Wilczek and Politzer.

Figure 8.1 displays a plot of the observed running of the strong coupling $\alpha_s = \frac{g_3^2}{4\pi}$ as a function of probed energy for a wide range of processes and energy scales and its comparison to the QCD prediction. The agreement with the experiment results is well established and the running of the QCD coupling is quite apparent.

The ramifications of asymptotic freedom were immediate and far reaching. Starting in 1968, SLAC began a series of experiments on the inelastic scattering of an electron from a proton: $e^-(\ell) + p(p) = e^-(\ell') + X$, where X can be anything. Here $\ell^\mu (\ell'^\mu)$ are the 4-momentum of the incident (final) electron and p^μ is the 4-momentum of the incident proton. Using general properties of Lorentz covariance and electric charge conservation, the differential cross for this inelastic process due to a single photon exchange in the rest frame of the incident proton where $p^\mu = (M, \vec{0})$, $\ell^\mu = (E, \vec{\ell})$, $\ell'^\mu = (E', \vec{\ell'})$ and θ is the scattering angle between the initial and final state electrons takes the form

$$\frac{d^2\sigma}{dE'd\Omega} = \frac{\alpha_{EM}^2}{4E^2 \sin^4\left(\frac{\theta}{2}\right)} \frac{1}{\nu} \left(F_2(x,q^2) \cos^2\left(\frac{\theta}{2}\right) + \frac{2\nu}{M} F_1(x,q^2) \sin^2\left(\frac{\theta}{2}\right) \right).$$

(8.8)

Here $q^\mu = (\ell' - \ell)^\mu$ is the momentum transfer, $\nu = \frac{qp}{M}$ and $0 < x = \frac{q^2}{2M\nu} < 1$. The $F_{1,2}(x,q^2)$ are the structure functions. In the rest frame, $\nu = E - E'$ gives the electron energy loss.

It was observed [63, 64] (for a review, see [65]) that in the deep inelastic limit (sometimes referred to as the Bjorken limit) where $q^2, \nu \to \infty$ at fixed x, the structure functions were independent of q^2 and dependent only on the dimensionless scaling variable x so $F_{1,2}(x,q^2) = F_{1,2}(x)$. This behavior was previously suggested by Bjorken [66] and is consistent with scattering off point-like constituents of the proton. Similar scaling behavior was also observed in deep inelastic muon- and neutrino-scattering experiments conducted at FNAL.

This scaling behavior could be accounted for by the parton model introduced by Feynman [67],[68] in his study of hadron–hadron collisions. The "naive" parton model describes the proton as a collection of point-like constituents called partons. At high momentum ("infinite momentum frame") the partons are taken as free. Therefore, the interaction of one parton with the electron does not affect the other partons and this leads to scaling in x. Today these partons are understood to be identical with the quarks postulated by Gell-Mann [69] in 1964 and independently by G. Zweig [70]. The quark-parton model was successfully applied by Bjorken and Paschos [71] to reproduce the observed scaling in deep inelastic electron-proton scattering. The observation of Bjorken scaling in deep inelastic scattering was instrumental in providing the first dynamical evidence of quarks and then gaining their acceptance as constituents of the proton. While the quark-parton model produced the scaling behavior observed at the time, the model itself lacked a firm theoretical foundation. It was not a quantum field theory and could not account for the strong interquark forces needed to account for the nonobservation of the partons in remnants of the collision. Due to its asympotitic freedom, QCD interaction strength falls at large momentum transfers as $\frac{1}{\ell n\, q^2}$ and thus behaves in a manner similar to the parton model. In fact, in subsequent experiments, deviations from the exact scaling behavior of the structure functions appeared. The scaling violations were precisely detected in a muon scattering experiment at FNAL in 1975 [72]. The observed q^2 dependence of the structure functions was accurately accounted for by QCD as reviewed in [73], marking a major success of the theory.

In a grander sense, the concept of asymptotic freedom provided a sensible interpretation of a quantum field theory at extreme ultraviolet energies. For nonasymptotically free theories, the running coupling increases with increasing energy diverging at the Landau pole[2] [74]. Asymptotic freedom in QCD also played an important role in theoretical work in early Universe cosmology. In the first few microseconds after the big bang, the Universe was comprised of a hot soup of quarks, leptons, gauge bosons and Higgs particles, which, as a consequence of asymptotic freedom, interacted very feebly and thus had properties amenable to calculation. This success in applying particle physics ideas to addressing puzzles in cosmology has led to a deep connection between the two disciplines, sometimes referred to as astroparticle physics, and has had a far-ranging influence on how one thinks about and treats cosmological problems. See [75] for a readable introduction. Asymptotic freedom has also been displayed in a variety of condensed matter systems. For a discussion of various such systems, see, for example, reference [151].

Motivated by the property of asymptotic freedom, it is suggestive that the attractive coupling grows at longer distances, leading to the confinement of all colored objects. Thus the quarks and gluons can never appear as physical states in the spectrum. Only net colored singlet hadrons (and glueballs) are physical states. Through the confinement mechanism, the potential long-range force that would naively arise from the massless gluons is obviated. This nonperturbative confinement mechanism has been demonstrated in lattice QCD [77], which is a well-established nonperturbative approach to solving QCD. The lattice gauge theory is formulated on a grid or lattice of points in space and time. When the size of the lattice is taken infinitely large and its sites infinitesimally close to each other, the continuum QCD is recovered. Additional details can be found in [78–80]. Lattice QCD has been used to compute the masses of the various hadrons as explicitly shown in reference [81].

In the limit of massless up and down quarks, QCD exhibits the continuous global $SU(2)_V \times SU(2)_A$ chiral symmetry, which is dynamically broken

---

[2] For pure QED with nonzero renormalized charge at the electron mass scale and using the 1-loop renormalization group running, one finds a Landau pole at 10^{227} GeV whose numerical value should certainly be questioned due to the approximation employed. For the weak hypercharge coupling in the SM model, using the analogous approximation, the Landau pole appears at 10^{34} GeV, which is still much larger than the Planck mass 10^{19} GeV.

to $SU(2)_V$ via the formation of a nonvanishing quark-antiquark vacuum condensate, which in turn gives mass to the nucleons. The pions are the pseudo Nambu–Goldstone bosons of the spontaneously broken axial $SU(2)_A$. (The pions develop a small mass as a consequence of the up and down quark mass, which provides a small explicit breaking of the $SU(2)_A$ symmetry). Thus QCD automatically displays the global symmetry structure of the strong interactions.

QCD has been very well tested and proven highly successful [62]. It constitutes one of the components of the Standard Model.

9 *SU*(2)$_L$ × *U*(1) Invariant Lepton Couplings to the Electroweak Vectors

As we have seen, simply adding a vector mass term to the Lagrangian violates the gauge invariance. However, in 1962, J. Schwinger argued that the gauge invariance might not preclude a vector mass once the radiative corrections are included, provided the current coupling to the vector is sufficiently strong that it can create a massless pole in the current-current correlation function, $\Pi(q^2)$. In that case, the full vector propagator varies as

$$\frac{1}{q^2}\frac{1}{1-\Pi(q^2)} \rightarrow \frac{1}{q^2}\frac{1}{1+\frac{M_V^2}{q^2}} = \frac{1}{q^2+M_V^2}, \tag{9.1}$$

which corresponds to a massive particle. The nonzero mass is said to be dynamically generated. While he was unable to demonstrate the existence of such a pole in any Lorentz-invariant model containing vectors in four space-time dimensions, he showed it was possible in a two space-time dimensional model, the Schwinger Model. In a very short, but extremely elegant paper [82], he studied a version of QED but in one space and one time dimension and with massless fermions. Since the fermions were massless, the QED$_2$ Lagrangian exhibits separate phase invariances for vector and axial vector transformations. Schwinger analytically solved the model and found a dynamically generated fermion-antifermion condensate that spontaneously breaks the axial symmetry and moreover the vacuum polarization function does indeed develop a massless pole so that the "photon" gets a nonvanishing mass. Although one can legitimately argue that the mass-generating mechanism is an artifact of the 1+1-dimensional nature of the model, this paper nonetheless was (and continues to be) extremely influential in that it showed that "vector" masses were possible in gauge theories.

A further elucidation of this idea was provided by P. Anderson [83] in 1963 who argued that the gauging of the spontaneously broken global continuous $U(1)$ symmetry in BCS theory essentially leads to a massive vector photon, which was then unable to penetrate the superconductor (Meissner effect [84]). Here the massless pole is provided by the massless phonons mediating between the Cooper pairs in the superconductor. The model provided a mechanism for vector mass generation in a nonrelativistic setting. Much discussion in the literature followed regarding the general applicability of Anderson's ideas to relativistic theories without any definitive resolution at the time.

Actually, there is an even earlier realization of the vector mass generation mechanism. In 1938, E. Stueckelberg [85] proposed a model of massive quantum electrodynamics using an additional scalar field that nonlinearly realized the symmetry. Like much of his work, it was well ahead of its time and largely ignored.

While the simple mimicking of QED by introducing vector bosons with direct mass terms was not an accurate description of the weak interactions, there was still considerable motivation for believing that the correct description should involve vector and axial vector fields. For one thing, the parity-violating extension of the Fermi theory was quite successful in describing low-energy, weak-interaction processes using lowest-order perturbation theory. A nice description of the phenomenological theory of the weak interactions before the advent of the electroweak gauge theory is contained in the monograph [86]. In addition, the Fermi description involved currents as did QED, albeit electric-charge-carrying currents that were both of a vector and an axial vector nature and moreover the hadronic piece of the vector current was conserved as is the vector current in QED. Finally, another property of the weak interactions where the resemblance to QED is striking is its universality [87],[88]. That is, the weak processes of β decay, muon decay and muon capture all had basically the same strength. This is reminiscent of the common magnitude of the electromagnetic charge of electrons, muons and protons.

J. Schwinger [40] was the first who attempted (unsuccessfully) to combine weak and electromagnetic interactions together in an $SU(2)$ Yang–Mills gauge theory using the charged massive $W^{\pm}$ and massless photons as the gauge vectors. His student S. Glashow [89] and independently A. Salam and J. Ward [90] did arrive at a model of leptons based on the gauge group $SU(2)_L \times U(1)$, which includes the correct couplings of the leptons to the electroweak gauge bosons. The model has two independent coupling constants g_2, g_1 associated with the $SU(2)_L$ and $U(1)$ gauge groups respectively. The product group has four generators and four associated vector bosons. To reproduce the V-A structure of charged weak interaction, the left-handed electron and its associated neutrino transform as a doublet, $L_L(x) = \begin{bmatrix} v_e(x) \\ e_L(x) \end{bmatrix}$, under $SU(2)_L$, while the right-handed electron, $e_R(x)$, is an $SU(2)_L$ singlet. Here we focus on the first generation of leptons consisting of the electron and its associated neutrino. The inclusion of the other lepton families is straightforward and will be discussed shortly. The model includes no right-handed neutrino field. To secure the couplings to the vector bosons, one follows the usual procedure of replacing $\partial_\mu L_L(x) \to D_\mu L_L$ and $\partial_\mu e_R(x) \to D_\mu e_R$ in the leptonic kinetic terms. Here the covariant derivatives are

$$D_\mu L_L(x) = \left(\partial_\mu - ig_2 \frac{\sigma_a}{2} W_{a_\mu}(x) + ig_1 \frac{1}{2} Y_\mu(x) \right) L_L(x),$$

$$D_\mu e_R(x) = \left(\partial_\mu + ig_1 Y_\mu(x) \right) e_R(x), \tag{9.2}$$

where σ_a ; $a = 1,2,3$ are the three Pauli matrices. The resultant locally invariant $SU(2)_L \times U(1)$ Lagrangian is

$$\mathcal{L}_e = -\bar{L}_L \gamma^\mu \frac{1}{i} D_\mu L_L - \bar{e}_R \gamma^\mu \frac{1}{i} D_\mu e_R. \tag{9.3}$$

The electron and its associated neutrino couplings to the gauge fields are thus dictated by the local $SU(2)_L \times U(1)$ invariance to be

$$\mathcal{L}_{int_e} = g_2 J^\mu_{a_e} W_{\mu a} + g_1 J^\mu_{Y_e} Y_\mu, \tag{9.4}$$

where the $SU(2)_L$ and $U(1)$ electron family currents are

$$J^\mu_{a_e} = \bar{L}_L \gamma^\mu \frac{\sigma_a}{2} L_L, \tag{9.5}$$

$$J^\mu_{Y_e} = -\frac{1}{2} \bar{L}_L \gamma^\mu L_L - \bar{e}_R \gamma^\mu e_R. \tag{9.6}$$

Two of the conserved currents of $SU(2)_L$,

$$J^\mu_{\pm_e} = 2(J^\mu_{1_e} \pm iJ^\mu_{2_e}), \tag{9.7}$$

are identified with the weak-interaction charged currents, which couple to the associated vectors

$$W^\mu_\pm = \frac{1}{\sqrt{2}} \left(W^\mu_1 \pm iW^\mu_2 \right) \tag{9.8}$$

as

$$\mathcal{L}_{cc_e} = \frac{g_2}{2\sqrt{2}} \left(J^\mu_{-_e} W_{\mu+} + J^\mu_{+_e} W_{\mu-} \right). \tag{9.9}$$

There are two neutral generators: the third component T_3 of the weak isospin $SU(2)_L$ and the weak hypercharge generator Y of the $U(1)$. One combination is identified as the electric charge generator

$$Q_{EM_e} = T_{3_e} + Y_e, \tag{9.10}$$

thus fixing the weak hypercharges with their values appearing in the covariant derivatives so that the electron has electric charge -1 and the neutrino is neutral.

The photon is identified as

$$A^\mu = \sin\theta_W W_3^\mu + \cos\theta_W Y^\mu, \tag{9.11}$$

which is the combination of vector bosons which couples to the electromagnetic current

$$J_{EM_e}^\mu = J_{3_e}^\mu + J_{Y_e}^\mu \tag{9.12}$$

as

$$\mathcal{L}_{EM_e} = e J_{EM_e}^\mu A_\mu. \tag{9.13}$$

Here θ_W, the Weinberg angle, is a weak mixing angle (the angle θ_W was originally introduced by Glashow as the weak mixing angle but has subsequently been universally referred to as the Weinberg angle), satisfying

$$e = g_1 \cos\theta_W = g_2 \sin\theta_W. \tag{9.14}$$

In addition, the model predicted a novel weak neutral current

$$J_{NC_e}^\mu = 2(J_{3_e}^\mu - \sin^2\theta_W J_{EM_e}^\mu), \tag{9.15}$$

which couples to a novel vector boson

$$Z^\mu = \cos\theta_W W_3^\mu - \sin\theta_W Y^\mu \tag{9.16}$$

as

$$\mathcal{L}_{NC_e} = \frac{e}{2\cos\theta_W \sin\theta_W} J_{NC_e}^\mu Z_\mu. \tag{9.17}$$

The muon and its associated neutrino and the tau and its neutrino have identical $SU(2)_L \times U(1)$ quantum numbers as the electron and its neutrino (see Table 9.1). Thus their invariant couplings to the electroweak gauge bosons have the same structure as the electron and its neutrino and the model exhibits a

Table 9.1 Weak isospin and weak hypercharge of the leptons.

Lepton	$SU(2)_L$	$U(1)$
$L_L = \begin{bmatrix} \nu_L \\ e_L \end{bmatrix}$	2	$-1/2$
e_R	1	-1

lepton universality. The full lepton-gauge interaction is then a sum of the three generations of leptons:

$$\mathcal{L}_{\text{int}} = \sum_{l=e,\mu,\tau} \left(e J^\mu_{EM_l} A_\mu + \frac{2}{2\sqrt{2}\sin\theta_W} (J^\mu_{-l} W_{\mu_+} + J^\mu_{+l} W_{\mu_-}) \right.$$

$$\left. + \frac{e}{2\cos\theta_W \sin\theta_W} J^\mu_{NC_l} Z_\mu \right). \tag{9.18}$$

The model introduced by Glashow, Salam and Ward is very incomplete:

- There is no consistent mechanism for generating the $W^\pm$ and Z vector masses.
- All leptons are massless, which is experimentally inaccurate.
- The model does not include quarks. The reason was that at the time the only known quark flavors were u, d, and s and the model extended to include them predicted far too large a rate for flavor (strangeness) changing neutral current (FCNC) processes such as $K_L \to \mu^+\mu^-$.

The model was successfully extended to include quarks in 1970 after the implementation of the S. Glashow, J. Iliopoulos, L. Maiani (GIM) mechanism [91], which introduced the c quark with the appropriate Cabibbo angle [93] mixing to provide a cancellation mechanism suppressing the dangerous FCNC processes. The c quark was subsequently discovered independently at the Brookhaven National Laboratory [94] and at SLAC [95] through the observation of the J/ψ meson. This mechanism was later extended by M. Kobayashi and T. Maskawa [96] to allow for the inclusion of the third-generation t and b quarks, which could further account for the observed CP-symmetry violation [92] in the electroweak interactions. The left-handed quark fields transform as doublets under $SU(2)_L$ while the right-handed quark fields are $SU(2)_L$ singlets. The weak hypercharge assignments lead to the electric charges $q_u = 2/3$, $q_d = -1/3$ as shown in Table 9.2.

Identical quantum numbers are assigned to the other generations of quarks by simply replacing (u, d) with (c, s) and (t, b).

Table 9.2 Weak isospin and weak
hypercharge of quarks.

Quark	$SU(2)_L$	$U(1)$
$Q_L = \begin{bmatrix} u_L \\ d_L \end{bmatrix}$	2	1/6
u_R	1	2/3
d_R	1	$-1/3$

10 Abelian Higgs Model

It was clear that the symmetry of the electroweak model had to be realized à la Nambu–Goldstone. What remained a major stumbling block was how to circumvent the theorem which mandated the presence of unobserved massless Nambu–Goldstone bosons. The solution was inadvertently arrived at in 1964 by R. Brout and F. Englert [97] and independently by P. Higgs [98]. Slightly later, G. Guralnik, C. R. Hagen and T. Kibble [99] also discussed in more detail how the massless NGB mode could be obviated. Another independent version of the mechanism was provided by A. Migdal and A. Polyakov [100]. (Although the time line of this publication is significantly behind the other work, it has been said that this was due to an initial rejection by the journal JETP to which it was submitted. No date has ever been substantiated for the original submission.)

The basic idea is to gauge the spontaneously broken global symmetry. The mechanism was actually initially presented in a simplified $U(1)$ toy model. Recall the global $U(1)$ symmetry of the Goldstone model and now promote it to a local Abelian U(1) symmetry, which yields the so-called Abelian Higgs Model. The model Lagrangian

$$\mathcal{L}_{Higgs}(x) = -(D_\mu \phi(x))^\dagger D^\mu \phi(x) - V(\phi) - \frac{1}{4}F^{\mu\nu}(x)F_{\mu\nu}(x)$$

with

$$D_\mu \phi(x) = \partial_\mu \phi(x) - iqA_\mu(x)\phi(x) \tag{10.1}$$

$$F_{\mu\nu}(x) = \partial_\mu A_\nu(x) - \partial_\nu A_\mu(x) \tag{10.2}$$

and

$$V(\phi) = \mu^2 \phi^\dagger \phi + \frac{1}{2}\lambda(\phi^\dagger \phi)^2 \tag{10.3}$$

is invariant under the local U(1) transformations

$$\phi(x) \to e^{iq\omega(x)}\phi(x)$$

$$A_\mu(x) \to A_\mu(x) + \partial_\mu \omega(x). \tag{10.4}$$

If $\mu^2 < 0$, the potential has the sombrero form displayed previously and is minimized for $\phi^\dagger \phi = \frac{v^2}{2}$ with $v^2 = -\frac{2\mu^2}{\lambda}$. Once again parametrizing

$$\phi(x) = \frac{1}{\sqrt{2}}(v + \phi_1(x) + i\phi_2(x)) \tag{10.5}$$

with Hermitian $\phi_{1,2}(x)$, the model Lagrangian (neglecting any field-independent terms) takes the form

$$\mathcal{L}_{Higgs} = -\frac{1}{2}(\partial_\mu \phi_1)^2 - \frac{(qv)^2}{2}\left(A_\mu + \frac{1}{qv}\partial_\mu \phi_2\right)^2 - \frac{1}{2}\lambda v^2 \phi_1^2 - \frac{1}{4}F^{\mu\nu}F_{\mu\nu} + \dots,$$

$$\tag{10.6}$$

with the ellipses denoting cubic and quartic interaction terms. Performing the gauge transformation

$$B_\mu(x) = A_\mu(x) + \frac{1}{qv}\partial_\mu \phi_2(x),$$

the second term is recognized as a mass term for the vector B_μ with mass $M_V = qv$. The scalar ϕ_1 is also massive, with $m_{\phi_1} = \sqrt{\lambda}v$. On the other hand, there is no kinetic term nor mass terms for ϕ_2, so its dynamics appears to have vanished. The apparently massless vector gauge field A^μ and the apparently massless scalar field ϕ_2 (the erstwhile Nambu–Goldstone mode) have combined to produce the massive vector field B^μ. The Nambu–Goldstone mode has become the longitudinal degree of freedom needed to provide the vector its mass. The number of degrees of freedom has not changed. In the vernacular, one says the NGB is eaten by the massless vector giving it mass. This is the celebrated Higgs mechanism. (Although it might be more properly called the Stuekelberg–Schwinger–Anderson–Brout–Englert–Higgs–Guralnik–Hagen–Kibble–Migdal–Polyakov–Weinberg–Salam–... mechanism.)

For every spontaneously broken global symmetry which is made locally (gauge) invariant, the erstwhile NGB is absorbed by the vector to become its longitudinal component, rendering the vector massive. This is true independent of the dynamics responsible for spontaneously breaking the global symmetry. The massive scalar mode (which in this toy model is the analog of the physical Higgs scalar) remains. The mechanism was extended by Brout, Englert and Higgs to non-Abelian models. Brout and Englert also mentioned the possibility of having a composite fermion-antifermion condensate be responsible for the symmetry breakdown. In all the papers, the motivation for the vector mass generation mechanism was that it be used in the strong interactions. None of the original authors foresaw its application to the weak interactions. Brout and Englert do not even mention the massive scalar mode in their 1964 paper. Higgs submitted two papers in 1964. The second paper was initially rejected, and the revised version does indeed mention the possible experimental implication of such a massive scalar in the spectrum. Guralnik, Hagen and Kibble also fail to discuss the massive scalar in their 1964 paper. So how is it possible to avoid the physical massless scalar degree of freedom when the spontaneously broken symmetry is gauged?

As alluded to previously, Goldstone, Salam and Weinberg [47] provided a proof of a theorem that mandates the emergence of a NGB mode in any theory each time the following three conditions are met:

(i) The appearance of a continuous global internal symmetry and corresponding conserved Noether current.

(ii) Said symmetry is spontaneously broken so there is a degenerate family of vacuum states and there exists a field (possibly composite) which carries a nontrivial charge of the spontaneously broken symmetry and whose vacuum expectation value is nontrivial.

(iii) The theory is manifestly Lorentz invariant.

Note that this last requirement is needed to guarantee the appearance of the Nambu–Goldstone bosons. While NGBs may also appear in theories with spontaneously broken global space-time symmetries or spontaneously broken global internal symmetries in Galilean invariant theories, they are not mandated to do so in all such cases while they are for manifestly Lorentz-invariant theories.

What fails in the case of gauge theories? To quantize a gauge theory, one must choose a gauge and impose a gauge-fixing condition. For the $U(1)$ gauge theory, a massless vector field has four components, but only two of them are physical (the transverse modes). One type of gauge choice, the Coulomb gauge defined by $\nabla \cdot \vec{A} = 0$ and $A^0 = 0$, explicitly eliminates the unphysical degrees of freedom. But the price one pays for this is the loss of manifest Lorentz invariance and an evasion of the theorem ensues. Alternatively, one could choose a manifestly Lorentz-covariant, gauge-fixing term such as a Landau gauge $\partial_\mu A^\mu = 0$. In this case, the Hilbert space necessarily contains the unphysical longitudinal and scalar degrees of freedom. Here the theorem does apply, but the contribution of the Nambu–Goldstone degrees of freedom to any physical process precisely cancels against the unphysical photon modes, so effectively these modes are uncoupled from the physical particles in the theory. Note that even though one must fix a gauge to quantize the model, the theory is very clever and all physical quantities are gauge and Lorentz invariant.

11 Standard Electroweak Model

In 1967, S. Weinberg [101] and A. Salam [102] independently developed a model of the electroweak interactions of leptons based on the gauge group $SU(2)_L \times U(1)$ using the Glashow and Salam–Ward model quantum number assignments and current identifications, but now exploiting the Higgs mechanism to give mass to the charged and neutral weak vector bosons while keeping the photon massless. Moreover, they exploited the Schwinger mechanism to generate the charged lepton masses. Explicitly, they introduced an $SU(2)_L$ doublet of scalar fields (four degrees of freedom), which also carries weak hypercharge $(-1/2)$. The scalars had a self-interacting potential function of the Brout–Englert, Higgs, Goldstone, Ginzburg–Landau form

$$V(\Phi) = \lambda \left(\Phi^\dagger \Phi - \frac{v^2}{2} \right)^2.$$

(11.1)

A nonzero vacuum expectation value, v, is generated for the neutral component of the scalar doublet spontaneously breaking the $SU(2)_L \times U(1)$ symmetry while leaving a residual unbroken electromagnetic symmetry $U(1)_{EM}$ so that $SU(2)_L \times U(1) \rightarrow U(1)_{EM}$. These scalars were coupled to the gauge fields via the covariant derivatives. The gauge vector of this unbroken symmetry was identified with the massless photon. The three erstwhile NGBs of the spontaneously broken symmetries became, via the Higgs et al. mechanism, the longitudinal components of the $W^\pm$ and Z vectors endowing them with masses, which satisfies

$$1 = \frac{M_W}{M_Z \cos \theta_W}.$$

(11.2)

(For details, the reader is directed to Appendix A.2.) Note that this relation follows as a consequence of using a scalar $SU(2)_L$ doublet to spontaneously break the symmetry. Let us focus on the preceding Higgs potential. Said potential actually exhibits a global $SU(2)_L \times SU(2)_R$ symmetry, which contains the global $SU(2)_L \times U(1)$ as a subgroup with the third generator of the $SU(2)_R$ identified with the weak hypercharge generator Y. This larger symmetry of the Higgs potential is not respected by the full electroweak Lagrangian, as it is explicitly broken by the gauging of the hypercharge and the differing Yukawa couplings between the upper and lower components of the $SU(2)_L$ fermion doublets so that the full Lagrangian has the $SU(2)_L \times U(1)$ symmetry. When the scalar doublet acquires its nontrivial VEV, the global $SU(2)_L \times SU(2)_R$ symmetry of the Higgs potential is spontaneously broken to a residual global $SU(2)$ symmetry. This residual $SU(2)$ symmetry is the diagonal subgroup of the $SU(2)_L \times SU(2)_R$ and is referred to as a custodial symmetry [103]. Note the unbroken electric charge generator, Q_{EM}, is the third component of the custodial $SU(2)$. If the spontaneous symmetry breaking is engendered by scalar fields transforming other than as doublets acquiring nontrivial VEV, the resulting Higgs potential will not exhibit the custodial $SU(2)$ symmetry and Eq. (11.2) will have no longer be satisfied.

The electroweak model has two independent gauge couplings whose values are extracted from experiment. The particular experiments employed to do so are often chosen to be those which are very accurately measured. One such choice is the electromagnetic fine structure constant. Two independent methods have been used. One uses the electron anomalous magnetic moment [62] to give $\alpha^{-1} = 137.035999150(33)$, while an independent determination [17] employs atomic interferometry in rubidium atoms to yield $\alpha^{-1} = 137.035999206(81)$.

Note that using the latter measure allows for a comparison between theory and experiment in the $g_e - 2$ determination. The second coupling is fixed by the Fermi constant, which is best secured today by comparing the calculation of the muon lifetime with the experimentally measured result [62] giving $G_F = 1.1663787(6) \times 10^{-5}$ GeV^{-2}. Thus while the theory is such that weak and electromagnetic interactions have a common origin, it is not technically a unified theory, which would have only a single coupling. The nontrivial VEV is then fixed by the Fermi scale as

$$v = (\sqrt{2}G_F)^{-1/2} \simeq 246.22 \text{ GeV}. \tag{11.3}$$

Very little attention was paid to the Glashow–Salam–Weinberg (GSW) model when it was first introduced. The turning point came in 1971, when G. 't Hooft [104] (see also [105–109]) proved that the model was renormalizable and hence consistent order by order in perturbation theory. The key was that the vector mass generation mechanism arose in a non-Abelian gauge theory using the Higgs mechanism. Since it is a Yang–Mills gauge theory, the vector propagator can be taken to have the same type of large momentum falloff as the photon propagator in QED even though the electroweak gauge bosons are massive.

A major breakthrough in the experimental verification of the model came with the discovery of the neutral current interactions in neutrino experiments using the Gargamelle bubble chamber at CERN in 1974 [110, 111]. The results of this experiment were consistent with the structure of the neutral current interactions as detailed in the GSW model.

In fact, long before the neutral current was observed, it was conjectured that they could exist and have parity-violating effects that could be detected [112] in both electron scattering experiments and via atomic spectra measurements. More than 20 years after this proposal, parity violation in the weak neutral current was indeed measured in polarized electron-deuteron scattering at SLAC [113, 114]. The agreement of this experiment with the predictions of the GSW model firmly established it as the leading candidate to describe electroweak interactions. The dominant contribution to the atomic bound state energy levels arises from the parity-conserving electromagnetic interaction between the electrons and the protons mediated by photon exchange. In addition, there will also be a contribution from the weak neutral current arising from the Z boson exchange. Since the neutral current interactions in the GSW model contain a parity-violating component, the interference term of these two contributions results in the various levels receiving a small admixture of the opposite parity. While an accurate quantitative estimate (for a general review see, for example, [115]) of this admixture requires detailed atomic physics calculations, for atomic number Z, the effect is roughly of order $\frac{Z^3 \alpha^2 m_e^2}{M_Z^2} \sim Z^3 \times 10^{-15}$. The Z^3 enhancement factor for large Z was first pointed out in [116]. Additional enhancement mechanisms can be employed to make the effect in cesium as large as 10^{-6}. The first such parity-violating signal was observed in 1978 by L. Barkov and M. Zolotorev [117] in Bi atoms. Over the following decades, parity violation has been observed in a variety of atoms, with the most accurate measurements made in Cs [118] by the Wieman group at Boulder.

In 1983, the $W^\pm$ and Z vector bosons were discovered by the UA1 and UA2 collaborations at CERN [119–122]. The world average of the measured masses is currently [62]

$$M_{W^\pm} = 80.379 \pm 0.012 \text{ GeV},$$
$$M_Z = 91.1875 \pm 0.0021 \text{ GeV}. \tag{11.4}$$

Using these measured masses, the weak mixing angle defined as $\sin^2 \theta_W = 1 - \frac{M_W^2}{M_Z^2}$ is determined as $\sin^2 \theta_W = 0.22337 \pm 0.00010$.

In Appendix A.2, the full $SU(2)_L \times U(1)$ electroweak model Lagrangian is presented in detail.

Tables 11.1 and 11.2 are taken from the particle data book [62] and provide a global fit to a variety of electroweak processes and the comparison to the Standard Model expectation. The agreement with the SM expectation is very good and the global electroweak fit describes the data very well. As of today, the electroweak model has been very precisely tested experimentally in a huge variety of different experiments covering a range of energy scales from the atomic level (eV) to the scale of the LHC (TeV) and its success constitutes a remarkable achievement.

One present potential outlier is the measurement of the anomalous magnetic moment of the muon $a_\mu = \frac{1}{2}(g_\mu - 2)$. The muon anomalous magnetic moment is a particularly promising quantity to investigate since any beyond-the-Standard-Model contribution scales like $a_\mu \sim (\frac{m_\mu}{M_{BSM}})^2$, where M_{BSM} is the mass scale of the beyond-the-Standard-Model physics. Hence, the sensitivity of a_μ is enhanced relative to a_e by a factor $(m_\mu/m_e)^2 \sim 4 \times 10^4$. Moreover, a_μ has been measured with a precision of 0.46 ppm [124], which is much more precise than what can currently be achieved experimentally for a_τ, which would, in principle, be an even more sensitive measure.

The SM expectation $a_\mu^{SM} = 116591810(43) \times 10^{-11}$ has been reported in a 2020 White Paper [123] by the Muon $g - 2$ Theory Initiative. The quoted overall precision of 0.37 ppm is limited by contributions due to the strong interaction, notably the contributions from hadronic vacuum polarization and to a lesser extent by the hadronic contribution to light-by-light scattering. This

Table 11.1 Standard Model prediction compared with experimental results secured using the LEP detector at CERN, which precisely measured various parameters in the e^+e^- collisions with total incident energy tuned to be near the Z mass. σ_{hadron} is total cross section to hadrons and Γ_Z is the total Z decay width. $R_l = \frac{\Gamma_{hadron}}{\Gamma_{l^+l^-}}$; $R_q = \frac{\Gamma_{q\bar{q}}}{\Gamma_{hadron}}$ for $l = e, \mu\tau$ and $q = b, c$ where Γ_{hadron} is the partial width for Z decays to hadrons and $\Gamma_{f\bar{f}}$ are the partial widths for $Z \to f\bar{f}$ with $f = e, \mu, \tau, u, d, s, c, b$; A_{FB}^f is the forward-backward asymmetry for fermion f; A_f is the left-right asymmetry for fermion f. Note that the three values for R_l are consistent with lepton universality. The column denoted by Pull gives the standard deviation from the SM prediction.

Quantity	Experiment	Standard Model	*Pull*
M_Z [GeV]	91.1876 ± 0.0021	91.1882 ± 0.0020	-0.3
Γ_Z [GeV]	2.4955 ± 0.0023	2.4942 ± 0.0009	0.6
σ_{hadron} [nb]	41.481 ± 0.033	41.482 ± 0.008	0.0
R_e	20.804 ± 0.050	20.736 ± 0.010	1.4
R_μ	20.784 ± 0.034	20.735 ± 0.010	1.4
R_τ	20.764 ± 0.045	20.781 ± 0.010	-0.4
R_b	0.21629 ± 0.00066	0.21581 ± 0.00002	0.7
R_c	0.1721 ± 0.0030	0.17221 ± 0.00003	0.0
A_{FB} (e)	0.0145 ± 0.0025	0.01619 ± 0.00007	-0.7
A_{FB}(b)	0.0996 ± 0.0016	0.1030 ± 0.0002	-2.1
A_{FB}(c)	0.0707 ± 0.0035	0.0736 ± 0.0002	-0.8
A_{FB} (s)	0.0976 ± 0.0114	0.1031 ± 0.0002	-0.5
A_e	0.15138 ± 0.00216	0.1469 ± 0.0003	2.1
A_b	0.923 ± 0.020	0.9347	-0.6
A_c	0.670 ± 0.027	0.6677 ± 0.0001	0.1
A_s	0.895 ± 0.091	0.9356	-0.4

recommended value for a_μ^{SM} is based on the "data-driven" evaluation of the hadronic vacuum polarization contribution, which is determined using dispersion relations and the experimentally measured hadronic cross sections in $e^+ - e^-$ collisions. Combining the run 1 data of the Fermilab Muon g-2 experiment [124–126] reported in 2021 and the previous measurement at the Brookhaven National Laboratory E821 Collaboration [127] results in a world average of $a_\mu^{expt} = 116592061(41) \times 10^{-11}$, which differs from the previously mentioned SM prediction of $a_\mu^{SM} = 116591810(43) \times 10^{-11}$ by $\Delta a_\mu = 251 \pm 59 \times 10^{-11}$, which constitutes a 4.2σ discrepancy. If this result persists,

Table 11.2 Standard Model prediction compared with experimental results for non-Z pole observables. m_t is the top quark mass, M_H is the Higgs boson mass, M_W is the $W^\pm$ mass and Γ_W is the $W^\pm$ width; $g^{\nu e}_{V,A}$ are the leptonic vector and leptonic axial vector couplings and τ is the lepton lifetime, τ_τ. The weak charges Q_W (Cs) and Q_W (Tl) are extracted from atomic physics parity violation experiments in Cs and Tl. The column denoted by Pull gives the standard deviation from the SM prediction.

Quantity	Experiment	Standard Model	*Pull*
m_t [GeV]	172.89 ± 0.59	173.19 ± 0.55	0.5
M_H [GeV]	125.30 ± 0.13	125.30 ± 0.13	0.0
M_W [GeV]	80.370 ± 0.019	80.361 ± 0.006	1.6
Γ_W [GeV]	2.195 ± 0.083	2.090 ± 0.001	-0.9
$g^{\nu e}_V$	-0.040 ± 0.015	-0.0398 ± 0.0001	0.0
$g^{\nu e}_A$	-0.507 ± 0.014	-0.5064	0.0
Q_W (Cs)	-72.82 ± 0.42	-73.23 ± 0.01	1.0
Q_W (Tl)	-116.4 ± 3.6	-116.88 ± 0.02	0.1
τ_τ [fs]	290.75 ± 0.36	288.90 ± 2.24	0.8

it indicates a violation of electron-muon universality beyond just the differences in their masses.

Lattice QCD provides an alternative theoretical calculational scheme to the data-driven methodology. Using an average of the lattice results available at the time, the Muon g-2 Theory Initiative White Paper [123] reported a higher value of a^{SM}_μ than that found using the data-driven method, but still consistent within the uncertainties. Subsequently, a lattice QCD result [128] for the hadronic vacuum polarization was published that reduced the difference between the SM and the experimental result to 1.5σ. They also estimated a $2 - 3\sigma$ tension in the hadronic contribution to the running of α than that determined using the $e^+ - e^-$ data. Clearly, an independent cross-check of the [128] lattice result is sorely needed.

Occasionally, other hints of deviations from the model have arisen. But thus far, none have stood the test of time.

A very nice demonstration of how the various elements of the Standard Model are required in order to have agreement with experiment is provided by the LEP measurement at CERN of the $e^+e^- \rightarrow W^+W^-$ cross section. At leading (second) order in perturbation theory, various Feynman graphs contribute: neutrino exchange, photon exchange and Z exchange, as shown in Figure 11.1. (There is also a Higgs exchange graph, which gives a contribution of $O(\frac{m_e}{v}) \sim 10^{-6}$ relative to the other graphs and can be safely ignored.)

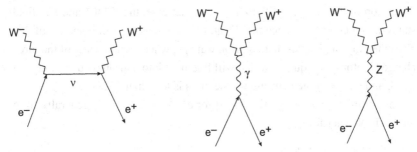

Figure 11.1 Leading-order Feynman graphs contributing to $e^+e^- \to W^+W^-$.

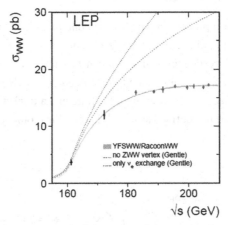

Figure 11.2 Measurements from LEP of the production cross-section $e^+e^- \to W^+W^-$ compared to the SM calculation of the RACOONWW [130], YFS [131] and GENTLE [132] collaborations. The plot is secured from the ALEPH, DELPHI, L3, OPAL and LEP Electroweak Collaborations [129]. Figure reproduced with permission from Ref. [129].

The neutrino exchange graph arises from the charged current interaction, while the photon exchange and Z exchanges mediate the electromagnetic and neutral current interactions respectively. Note that the Z exchange graph involves the triple gauge vector ZW^+W^- vertex, which is mandated by the non-Abelian gauge structure. Plotted in Figure 11.2 is the LEP measurement as well as the predicted cross section with only neutrino exchange, no ZWW vertex and with all three as required by the Standard Model. Clearly all three interactions are required for agreement.

In the SM, the Z vector boson couples directly via the neutral current interaction to $f\bar{f}$ where f is any fermion and to W^+W^-. It also couples to HH with H the Higgs boson. Note there is no direct coupling of the Z to a pair of photons. Using the LEP detector at CERN, the e^+e^- beam energy was tuned to scan the region in the vicinity of the Z mass. In that case, the W^+W^- and HH final states do not contribute since the $W^\pm$ and H have masses well in excess of $\frac{M_Z}{2}$. Thus the only final states that contribute are $f\bar{f}$, where f can be any of the three charged leptons, any quark (which will fragment into hadrons) other than the t quark and any possible neutrino whose mass is less than $\frac{M_Z}{2}$.

The annihilation cross section [133] for $e^+e^- \to Z \to f\bar{f}$ is generally fit to the Breit–Wigner form

$$\sigma_f = \frac{12\pi}{M_Z^2} \frac{s\Gamma_e\Gamma_f}{(s - M_Z^2)^2 + s^2\frac{\Gamma_Z^2}{M_Z^2}} \quad ; \quad s \simeq M_Z, \tag{11.5}$$

which accounts for a process in which a spin one particle is produced in e^+e^- collisions and then annihilates into an $f\bar{f}$ channel in the resonance region for the case where the resonance width is much less than the vector mass. This typical resonance shape peaks around the Z mass, $\sqrt{s} = M_Z$, and has a total width Γ_Z. If the Z decays a fraction B_f of the time into a final state $f\bar{f}$, the corresponding partial width is defined as $\Gamma_f = B_f\Gamma_Z$. For the moment, we assume that

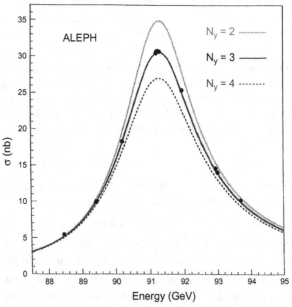

Figure 11.3 The e^+e^- annihilation cross section as a function of center-of-mass energy in the vicinity of the Z pole, as measured by the LEP experiments [133]. The curves represent the Standard Model predictions for two, three and four species of light neutrinos. It is clear from this picture that there is no further light neutrino species with couplings identical to the first three.

there are N_ν neutrino species that contribute. In that case, the total Z width is simply

$$\Gamma_Z = 3\Gamma_e + \Gamma_{hadron} + N_\nu \Gamma_\nu. \tag{11.6}$$

The LEP detector separately measured the total width Γ_Z from the shape of the resonance curve as well as the hadronic width, Γ_{hadron} ($\sim 0.7 \ \Gamma_Z$) and each charged leptonic width Γ_e ($\sim 0.03 \ \Gamma_Z$). Using the Standard Model calculation of Γ_ν, the only unknown is N_ν, the number of light neutrino species. Figure 11.3 displays the line shape for $N_\nu = 2, 3, 4$. The fit to the data yields $N_\nu = 3.0026 \pm 0.0061$, which is in excellent agreement with the observed number of fermion generations: $N = 3$.

12 The Higgs Boson

By 2012, the only as yet undetected ingredient of the Standard Model was the Higgs scalar H. As detailed in Appendix A.2, it couples to all massive bosonic particles in proportion to the square of the boson mass and to fermionic particles linearly in the fermion mass. Furthermore, the generalized Yukawa couplings to the fermions also allow for the observed flavor-changing charged weak interactions parametrized by the Cabibbo–Kobayashi–Maskawa (CKM) matrix [93, 96]. (For the experimentally determined CKM parameters, see [62].) The Higgs scalar is self-coupled and massive $M_H^2 = \frac{1}{2}\lambda v^2$. The model, however, does not fix the value of this mass, which must be determined experimentally. This was achieved in 2012 when the CMS and Atlas collaborations at the CERN LHC announced the discovery [1, 2] of a scalar particle with mass ~ 125 GeV whose properties were consistent with that of the Standard Model Higgs boson.

The dominant decay channel of a Higgs scalar with such a mass is into $b\bar{b}$ with a branching fraction [62]: $BR(H \to b\bar{b}) = 0.58$. However, at hadron colliders such as the LHC, this mode is masked by the presence of very large backgrounds, making the identification of a Higgs boson signal in these channels quite challenging. Hence the CMS and Atlas collaborations focused on the decay of the Higgs scalar into two photons: $H \to \gamma\gamma$. Since the photon is massless, there is no direct coupling to the Higgs field. The leading contribution to this process comes from the 1-loop diagrams, with the more massive charged particles traversing the loop giving the dominant contributions. This corresponds to the top quark and the W boson as shown in Figure 12.1.

Consequently, the branching fraction [62] for this mode is small: $BR(H \to \gamma\gamma) = 2.27 \times 10^{-3}$. Nonetheless, an unambiguous signal was successfully detected. Subsequently, additional decay channels including $b\bar{b}$ have been

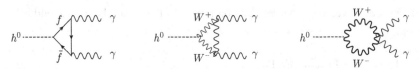

Figure 12.1 Examples of Feynman graphs contributing to $H \to \gamma\gamma$. In the first diagram, f can be any charged, massive fermion with the dominant contribution coming from the top quark traversing the loop.

observed and analyzed. In all cases, consistency with the properties of the Standard Model Higgs boson persists. As of this writing, the mass of this scalar is measured [62] as

$$M_H = 125.30 \pm 0.09 \pm 0.09 \text{ GeV}. \tag{12.1}$$

A few comments are in order.

- One really needs to distinguish between the existence of the Higgs et al. mechanism and the particular way it is engendered. The Higgs mechanism that gives mass to the weak interaction vector bosons results whenever a spontaneously broken global continuous symmetry is made locally invariant independent of the dynamics responsible for the spontaneous symmetry breakdown. It was actually validated in 1983 with the discovery of the massive $W^\pm$ and Z vector bosons. With the discovery of a light Higgs scalar particle, it appears this mechanism arises in the electroweak model through the interactions of a perturbatively coupled scalar field sector based on the type proposed by Brout–Englert and Higgs and others. The physical dynamics implementing the vector mass generation mechanism did not necessarily have to be of this form. For example, it could have been some new interaction leading to a fermion-antifermion condensate breaking the electroweak symmetry more in analogy to what happens in BCS or to the chiral symmetry breaking in QCD. In that case, the "Higgs scalar" would not be nearly as light as it turned out.
- The Higgs mechanism as implemented in the Standard Electroweak Model is extremely economical. With the addition of one novel scalar physical degree of freedom, one generates nonzero masses for the weak gauge vectors and the fermions as well as incorporates the observed Cabibbo–Kobayashi–Maskawa quark flavor mixings and CP-symmetry violation in a mathematically consistent framework. Nature apparently has chosen the highly economical and simplest, albeit arguably not the most satisfying route. While all the rest of the dynamics in the Standard Model is determined via interactions with the gauge bosons, the scalar self-couplings responsible for

the symmetry breaking and the Yukawa couplings to the scalar fields are not.

- Contrary to what sometimes appears in the literature, the Higgs mechanism is not responsible for most of the visible (i.e. nondark matter) mass in the Universe. The predominance of visible mass is protons and neutrons. Most of their mass results from the QCD binding of their quark and gluon constituents. Up and down quark masses do arise from the Higgs mechanism, but their contribution to the proton mass is very small (a few parts in a thousand). However, the fermion masses generated via the Higgs/Schwinger mechanism are crucial. Without it, QCD still confines quarks into protons so the nucleon mass would be basically unchanged. However, now the electroweak symmetry would be broken by the formation of QCD vacuum chiral condensates at a much lower scale ($\sim$250 MeV). The W and Z bosons would acquire mass of this order by absorbing the Nambu–Goldstone boson pions. Because the electron would be massless in the absence of the Higgs/Schwinger mechanism, the atomic Bohr radius would be infinite. There would be nothing recognizable as an atom, no chemistry as we know it, no stable composite structures like solids or liquids. None of us would exist!

13 Global Symmetries versus Local Invariance

We have traced the development of the Standard Model of particle physics by following how its various symmetries were uncovered and implemented. While recognizing that global symmetries played a crucial role in this story, as it turns out, the global symmetry structure is actually an automatic consequence of the local invariance. That is, once the gauge group and transformation properties of the matter fields under said group are specified, the dynamics is determined and so is the global symmetry structure, which does not have be separately mandated and enforced.

The case of gauge invariance is somewhat more subtle. Calling an invariance under local transformations a symmetry (gauge symmetry) is actually a misnomer. When one speaks of a symmetry in a quantum theory, what one really means is the invariance under transformations of the quantum states that lie in a Hilbert space. However, a gauge transformation does not change the quantum states (it acts trivially on the Hilbert space) and hence it really should not be called a symmetry. Rather the gauge invariance dictates that the excitations of certain fields appearing in the action functional do not really produce physical states. Thus the gauge symmetry actually corresponds to a redundancy of the theoretical description, which is introduced to insure that the theory is local. This is why the gauge invariance of the underlying action functional must be exact[3] and cannot be broken by the inclusion of additional non–gauge-invariant local operators without destroying the consistency of theory. With the inclusion of such non–gauge-invariant terms in the action, the unphysical degrees of freedom would not be eliminated, leading to a violation of some basic principle such as Lorentz invariance or positivity of the state space norm, which corresponds to a breakdown of causality or unitarity (probabilities exceeding unity).

The same argument applies to the case where one attempts to gauge a symmetry of the Lagrangian which is broken by quantum radiative corrections arising from fermion loops. In such a case, the symmetry is said to be anomalous. Anomalies [134, 135] reflect an intrinsic breaking, which cannot be compensated by adding local counterterms in higher orders of perturbation theory. Thus any potential anomalies associated with the dynamical gauge currents need to be canceled if the dynamical gauge invariance is to be preserved.

Since the anomalies depend only on the charge structure of the dynamical fermions traversing the loops, their cancellation constrains the fermion matter content of many gauge field theories. The nonrenormalization theorem [136] guarantees that this cancellation will be preserved to all orders. From the form of the non-Abelian anomaly [137], it can be shown that models with vector-like gauge couplings, such as QED or QCD, do not have dynamical anomalies. Only theories where the fermions have chiral gauge couplings can have nontrivial anomalies. If the standard electroweak model is based on a chiral gauge theory, the individual fermions propagating in the loop produce anomalies, but are canceled between the various quark and lepton contributions.

This is contrary to the case of global symmetries where globally nonvariant local terms can be added to the action without destroying the consistency of the theory. The only way to "break" a gauge symmetry is by making the spontaneously broken global theory locally invariant. But this operation is not a deformation of the theory (as is the case of adding globally noninvariant terms to an otherwise globally invariant theory), but rather corresponds to a different theory altogether. There is also no restriction on having anomalous global symmetries. In fact, the existence of an anomaly in the axial vector global current is responsible for the experimentally observed decay mode of the neutral pion to two photons.

---

[3] Strictly speaking, it is the classical action that must be gauge invariant. To quantize the theory, a gauge must be chosen that does indeed break the gauge invariance. However, said gauge fixing terms (and accompanying Faddeev–Popov ghost terms) are of just such a nature that all physical observables are gauge invariant. See Appendix A.2 for some additional discussion.

14 Limitations and Challenges of the Standard Model

Despite being the most successful theory of particle physics to date, the Standard Model is still neither complete nor totally satisfying. Here we briefly address shortcomings of the model.

• Standard Model parameters: Various parameters appear in the Standard Model. The form of the gauge groups and the quantum numbers of the fermions and scalars under said gauge groups are taken as given a priori. The same can be said of the 3+1 dimensionality of space-time. In addition, as in any quantum field theory, the Lagrangian must be supplemented by various normalization conditions before it is completely defined and used to make unambiguous predictions. The Standard Model depends on 19 parameters, each of which requires a normalization condition that can be fixed by experiment. These include the three gauge couplings, the three charged lepton masses, the six quark flavor masses, the three CKM mixing angles, one CKM phase and two parameters from the scalar sector that can be taken as the Higgs boson mass and the electroweak scale. In the Standard Model, only massless left-handed neutrinos appear.

A variety of beyond-the-Standard-Model proposals have been forwarded in an attempt to reduce the number of said parameters. Thus far, none have successfully achieved this goal. One approach posits the existence of a unified gauge group with a single gauge coupling that spontaneously breaks at a very high energy scale to the Standard Mode direct product gauge groups. While these Grand Unified Theories (GUTS) replace the three gauge couplings of the SM with one unified gauge coupling, all such models need to introduce a plethora of new particles and a large number of additional couplings both to achieve the unification and to engender the appropriate spontaneous symmetry breaking. Thus far, none of these have been observed. In addition, one experimental consequence of many GUTS is the finite lifetime of the proton and, as yet, all experimental searches have yielded null results. A review of GUTS is provided in [138].

In addition, there is another hitherto neglected parameter in the QCD Lagrangian, θ, which is the coefficient of the gauge-invariant Lagrangian term

$$\mathcal{L}_\theta = \theta \frac{g_3^2}{32\pi^2} G_i^{\mu\nu}(x)\tilde{G}_{\mu\nu\,i}(x) \tag{14.1}$$

where $G_i^{\mu\nu}$ is the gluon field strength and

$$\tilde{G}_{\mu\nu\,i} = \frac{1}{2}\epsilon_{\mu\nu\lambda\rho}G_i^{\lambda\rho} \tag{14.2}$$

is the field strength dual. This gauge-invariant CP-symmetry-violating product is a total divergence [140],

$$\mathcal{L}_\theta = \theta \frac{g_3^2}{32\pi^2} \partial_\mu K^\mu(x), \tag{14.3}$$

of the gauge-variant combination

$$K^\mu = \epsilon^{\mu\alpha\beta\gamma} A_{\alpha i} \left(G_{\beta\gamma \, i} - \frac{g_3}{3} f_{ijk} A_{\beta \, j} A_{\gamma \, k} \right). \tag{14.4}$$

Being a total divergence, $\mathcal{L}_\theta$ never contributes in any perturbative calculation. However, the QCD vacuum has a nontrivial structure, and nonperturbative effects can become important. Experimentally, the size of θ is constrained by measurements of the CP-symmetry-violating electron dipole moment (EDM) of the neutron. So far, no neutron EDM has been observed and the current experimental limit [139] is $d_n < (0.0 \pm 1.1) \times 10^{-26}$ e cm. The violation of CP-symmetry in the SM through the CKM phase gives $|d_n| \sim 10^{-31}$ e cm, which is far smaller than the error bars on the current bound. Since the θ parameter provides another possible source of CP-symmetry violation, the measurement of the EDM of the neutron restricts it to satisfy $\theta < 10^{-10}$. The origin of the smallness of this dimensionless parameter is referred to as the strong CP problem. One possible solution is afforded by the Peccei–Quinn mechanism [141],[142], which introduces some additional degrees of freedom and an additional $U(1)_{PQ}$ spontaneously broken anomalous symmetry, which allows this parameter to be effectively removed. There is a residual effect in the form of a novel pseudoscalar particle, the axion [143],[144], which is actively being hunted experimentally [62]. The pseudo-NGB axion mass is model dependent but is roughly given by $m_a \sim \frac{m_{EB}^2}{f_a}$, where f_a is the scale associated with the $U(1)_{PQ}$ spontaneous breaking and m_{EB} is the mass scale set by the explicit $U(1)_{PQ}$ breaking. For so-called QCD axions, $m_{EB} \sim \langle \bar{\psi}\psi \rangle^{1/3} \sim 250$ MeV, where $\langle \bar{\psi}\psi \rangle$ is the QCD chiral symmetry-breaking condensate and provides the mass scale associated with the chiral anomaly in QCD. To avoid adversely impacting stellar cooling rates, there are stringent constraints (see, for example, [147–149]) highly suppressing axion–SM interactions. Allowed models tend to favor a small axion mass ($\leq 10^{-3}$ eV) whose couplings to SM particles vary as $f_a^{-1} \leq 10^{-10}$ GeV $^{-1}$ and hence are very feeble. If axions exist, they are a possible component of cold dark matter.

• Neutrinos: In the Standard Model, neutrinos are massless particles. However, neutrino oscillation experiments have shown that neutrinos do have nonzero masses [145, 146]. The possibility of neutrino oscillations was first proposed by B. Pontecorvo [150] in 1957, in analogy with neutral kaon mixing. Nonvanishing neutrino masses can be accounted for in a variety of ways by suitably modifying the Standard Model. The actual mechanism(s) of this mass generation is still unknown and constitutes a very active area of investigation.

For spin 1/2 fermions, Lorentz invariance allows two type of mass terms. A Dirac mass couples the left-handed and right-handed components of the fermion. In the Standard Model, the charged lepton and quark masses are all of this type. It is also possible, however, to have a Lorentz-invariant mass term using only a single handedness. This requires that said fermion field be self-conjugate so that its particle excitations are their own antiparticles. As such, so long as electric charge is conserved, only electrically neutral particles can have this property and neutrinos are possible candidates. The possibility of such a mass term for fermions was introduced in 1937 by E. Majorana in his last published work [151] and they are referred to as Majorana fermions.

Only left-handed neutrinos have thus far been observed experimentally. If one insists on using only Standard Model particles, which has only left-handed neutrinos, the only possibility is for these left-handed neutrinos to have Majorana masses. This can be achieved by adding a nonrenormalizable mass dimension-five interaction of the lepton doublet with the scalar doublet bilinear [152]. However, since the neutrino masses are considerably smaller than the rest of the known particles (at least 500,000 times smaller than the mass of an electron), it follows that the coefficient of this operator is the inverse of a very large mass scale, which must originate from outside the SM.

Another possible mechanism involves the introduction of right-handed neutrino fields as $SU(2)_L \times U(1)$ singlets. They can then be coupled to the left-handed lepton doublet by introducing additional Yukawa couplings to the Standard Model scalar doublet. Following an analogous procedure to the CKM mixing in the charged current quark interactions (see Appendix A.2) leads to neutrinos with Dirac masses as well as the Maki–Nakagawa–Sakata (MNS) mixing matrix [153] in the lepton charged current interaction. (For the experimentally allowed range of MNS parameters, see [62].) Once again, because the observed neutrino masses are so much lighter than all the other fermions, use of the Higgs mechanism to generate their masses requires extremely small Yukawa couplings, which is not particularly attractive.

Currently, a popular conjectured approach for generating very light neutrino masses is the seesaw mechanism [154, 155], where the right-handed neutrinos have very large Majorana mass terms. Note that this new mass scale is allowable, since the right-handed neutrinos are $SU(2)_L \times U(1)$ singlets. These right-handed neutrinos can also couple to the left-handed lepton doublets through the Yukawa interactions with the scalar doublet. Assuming these Yukawa couplings are of the order unity, then after the electroweak spontaneous symmetry breaking and diagonalization of the neutrino mass matrix, one finds two sets of massive Majorana neutrinos. One set have very large masses and are composed mostly of the right-handed neutrinos with a small admixture of the left-handed ones, while the other have masses inversely proportional to the heavy mass parameters and hence are very light. These are composed mainly of the original left-handed neutrinos with a small admixture of the right-handed neutrinos. To obtain the correct order of magnitude of the observed light neutrino masses, the Majorana mass scale of the right-handed neutrinos is required to be around 10^{14} GeV.

• Dark matter: The hypothesis of the necessity of dark matter (DM) in the Universe has a long history (For a discussion of the early history of dark matter proposals, see, for example, [156]), including a talk given by Lord Kelvin in 1884 in which he stated his estimate of the mass of the visible stars in the Milky Way was insufficient to account for the observed velocity dispersion of the stars orbiting around the center of the galaxy. Further indications came in the 1930s by F. Zwicky's observation that the mass-to-light ratio was not unity from measurements of galaxy rotation curves [157, 158]. While the numerical values of the estimates were more than an order of magnitude inaccurate, mainly due to an incorrect value of the Hubble constant, they did lead to the correct conclusion that the bulk of the matter was dark.

By the 1980s, studies of these galactic rotation rotation curves [159, 160] provided stronger evidence for the existence of dark matter. In addition, more recently there have been a plethora of observations substantiating the hypothesis. These include the analysis of the power spectrum of the cosmic microwave background (CMB) [161], which indicates that about 24 percent of the energy density present in the Universe is in the form of dark matter. For a review, see, for example, [162]. Additional evidence for dark matter comes from observations of gravitational lensing [163] along with astronomical observations of the motion of galaxies within galaxy (bullet) clusters [164, 165]. Dark matter also plays a crucial role in structure formation because it predominantly feels only the force of gravity. Thus, the gravitational Jeans instability, which allows compact structures to form, is not countered by any other interaction, such as radiation pressure. Consequently, dark matter begins to collapse into a complex network of dark matter halos well before ordinary matter, which is impeded by the pressure forces. Without dark matter, the epoch of galaxy formation would occur substantially later in the Universe than what is observed [166, 167]. As such, it has became clear that the bulk of matter holding galaxies together is in extended halos of nonluminous dark matter.

To date, dark matter has been observed only astrophysically via its gravitational interaction and has eluded any direct laboratory detection. Thus it is a form of matter that interacts very feebly with the Standard Model fields, in particular with the electromagnetic field. The Standard Model itself does not supply any good dark matter candidates. Cold dark matter (CDM) [169, 170] offers the simplest explanation for most cosmological observations. Cold refers to the fact that the dark matter moves slowly compared to the speed of light. It is dark matter composed of constituents with a free streaming length much smaller than a protogalaxy. This is the focus for most dark matter research, as hot dark matter does not seem capable of supporting the observed galaxy or galaxy cluster formation. At present, a variety of experiments probing the so-called direct, indirect and collider channels are underway [168]. The main focus of these searches has been to identify cold dark matter as weakly interacting massive particles (WIMPs). The appeal of WIMP dark matter is due in part to the suggestive coincidence between the thermal abundance of WIMPs and the observed dark matter density (WIMP miracle). Many particle physics theories beyond the Standard Model provide natural candidates for WIMPs and there is a huge range in the possible WIMP masses (1 GeV to 100 TeV) and interaction cross sections with normal matter (10^{-40} cm^2 to 10^{-50} cm^2). As the allowed parameter space for WIMPs continues to shrink, there has been growing attention to other, generally much lighter (sub GeV), dark matter candidates. In general, since lighter DM particles have less available kinetic energy, achieving a kinematic match between the DM and the target requires a proper treatment of collective excitations such as charged quasi-particles and phonons in the condensed matter systems, hence necessitating an interdisciplinary approach. While most such light DM proposals lack specific well-motivated candidates nor have the appeal provided by the WIMP miracle, there is one very light DM candidate in the form of the axion whose existence is independently motivated by the Peccei–Quinn solution to the strong CP problem.

• Dark energy: In physical cosmology and astronomy, dark energy is an unknown form of energy that affects the Universe on the largest scales. (For reviews, see, for example, [171, 172].) The first direct evidence for its existence came in 1998 from the High-Z Supernova Search Team observations [173] of Type Ia supernovae. In 1999, the Supernova Cosmology Project [174, 175] followed by suggesting that the expansion of the Universe is accelerating. Using observed data, the first paper that made this claim was [176]. Without introducing a new form of energy, there was no way to explain how such an accelerating Universe could arise. Since the 1990s, dark energy has been the most accepted premise to account for the accelerated expansion. The best current measurements indicate that dark energy contributes 71 percent of the total energy in the present-day observable Universe. Note that for a flat Universe with critical density given by WMAP [161] and Planck [177, 178], this corresponds to a dark energy density of only $(10^{-12}$ GeV$)^4 \sim 2 \times 10^{-10}$ erg/cm^3, which is much less than the density of ordinary matter or dark matter within galaxies. However, it dominates the energy of the Universe because it is uniform across space. The introduction of a cosmological constant, Λ, [179] into the Einstein action is one way to account for dark energy, and the ΛCDM model is supported by a wealth of cosmological data.

The energy density of the vacuum is notoriously difficult to calculate in quantum field theory [180]. There are at least two distinct contributions to said vacuum energy which arise in the SM during the evolution of the Universe. In the standard electroweak model, the phases of broken and unbroken symmetry are distinguished by a potential energy difference of approximately ~250 GeV (where 1 GeV = 1.60 $\times 10^{-3}$ erg = 5.06 $\times 10^{13}$ cm^{-1}). The Universe is in the broken-symmetry phase during our current low-temperature epoch, and is believed to have been in the symmetric phase at sufficiently high temperatures at early times. Unless there is a very fine tuning, the effective cosmological constant is therefore different in the two epochs and one would naturally expect a contribution to the vacuum energy density today of order ~(250 GeV)^4 ~ 7.3×10^{47} erg/cm^3. Similarly, another contribution can arise due to quark-antiquark vacuum condensate formation in QCD, which spontaneously breaks the chiral symmetry breaking: $\langle |q\bar{q}|0\rangle$ ~ $(0.25 \times 10 \text{ GeV})^3$. The associated contribution to a vacuum energy density is then expected to be ~(0.25 GeV)^4 ~ 7.2×10^{35} erg/cm^3. There could be even larger contributions coming from unknown transitions occurring earlier in the Universe's evolution. The ratio of the energy scale (the energy density to the 1/4 power) of these contributions differs by some 12 to 14 orders of magnitude from the observed cosmological constant scale. Clearly this difference constitutes a (cosmological constant) problem of enormous import. In principle, one could impose an extreme fine-tuning at a very high scale (before any of the transitions) where the unknown quantum theory of gravity would be operational, to miraculously cancel these various contributions. This is not a very satisfactory explanation, as the various contributions arise at different scales and have a different physical origin.

• Naturalness: Naturalness is a concept that has emerged in recent years as a discriminator of various theoretical models. It has taken on a variety of different meanings. Crudely, the notion of naturalness is that one prefers not to have to fine-tune the various parameters in a model in order to generate a physical observable that is much smaller than it would otherwise be. In principle, there is nothing wrong with such a fine-tuning, as it does not violate any physical principle or tenet. It is just that such a fine-tuning leaves most physicists uncomfortable and they would prefer models where it is unnecessary. One example of a naturalness problem is the value of the cosmological constant discussed previously.

Here we focus on the SM Higgs boson mass. Quantum loop corrections can affect the value of this mass. For example, the 1-loop top quark graph provides an $O(m_t^2)$ contribution to the Higgs mass. Assuming the SM gives an accurate description of the physics up to a large energy scale, it is often stated that the quadratic divergences of the loop amplitude shift the Higgs mass by a large amount. The fine-tuning of the bare mass term required to keep the physical Higgs mass at its experimentally measured value and not at the high mass scale then constitutes a naturalness problem in the SM. A more precise definition of naturalness due to G. 't Hooft [196] is that a parameter is natural if by setting it to zero, one increases the symmetry of the model, which, when enforced, keeps the parameter zero. For the case of the naturalness of the Higgs mass, the most commonly invoked symmetry is supersymmetry (SUSY). (For an introduction, see [197].) Here one adds additional degrees of freedom to the model such that there are equal numbers of bosonic and fermionic fields and moreover the couplings of these fields are related in a well-defined way. If such a supersymmetry exists, then the quadratic divergence of the Higgs mass due to a Fermi field propagating the loop is identically canceled by a quadratic divergence arising from a loop of its bosonic partner field. The relative minus sign is a reflection of the opposite statistics. Thus in supersymmetric theories, quantum corrections do not renormalize the superpotential that contains the Higgs mass term. Of course, SUSY is not an exact symmetry of nature and it must be broken. In these models, the electroweak symmetry breaking is generated by terms that are a soft, explicit breaking of the SUSY. As such, as long as the SUSY breaking scale is not too large, the Higgs mass is protected.

Rather than introducing all the additional, as yet unobserved, degrees of freedom and new interactions required to produce a softly broken SUSY extension of the SM, the standard electroweak model Lagrangian already possesses a scale symmetry broken only by the negative squared mass term for the scalar doublet which can act as a protection mechanism of the Higgs mass [198]. Just as in the SUSY case, soft symmetry breaking will set the scale of the electroweak symmetry breaking and generate the Higgs mass. This mechanism is completely consistent at tree level. Perturbative loop corrections will introduce an explicit scale symmetry breaking through the scale anomaly related to the logarithmic running of the couplings. Since the quadratic divergences are unrelated to the coupling runnings, they represent a separate explicit breaking of the tree-level scale symmetry which is generated by the introduction of an explicit momentum space cutoff procedure. In other cases where the cutoff procedure breaks a global symmetry, counter-terms are added to restore the original symmetry structure of the theory. This does not constitute a fine-tuning issue, but merely is an artifact of the specific computational scheme used to regularize the loop amplitudes. One can make a similar argument that the quadratic divergences in the SM are the spurious effects of a particular regularization scheme. Counter-terms must be added to preserve the structure of the anomalous scale symmetry Ward identity. This argument can be used to remove the explicit quadratic divergences order by order in perturbation theory without the need to fine-tune. Note that the usual dimensional regularization procedure would generate only the logarithmic running of the couplings and not quadratic divergences.

Of course, it is still possible that a fine-tuning might be required due to nonperturbative effects such as the presence of a Landau pole, which appears in the scalar self-coupling and weak hypercharge gauge coupling or when the SM is embedded in a more all-encompassing model, which could have physical mass scales at far-ultraviolet energies. Whether scale invariance can still be used to protect the electroweak scale depends on the scaling properties of the more underlying model. Of course, the softly broken scale symmetry does not explain the origin of the electroweak scale, just as is the case in SUSY extensions where the SUSY breaking scale must be adjusted by hand to produce the correct electroweak scale.

• Matter-antimatter asymmetry or Baryogenesis: The Big Bang should have created equal amounts of matter and antimatter in the early Universe. Today, however, all nondark matter observed is made almost entirely of matter or baryons. There is comparatively very little antimatter or antibaryons to be found. The observed abundance of baryons today implies that when the universe was much hotter than a GeV the ratio of antibaryons to baryons must have been about one part in 10^8. Something must have happened to create this asymmetry. (For a review, see, for example, [181].) In 1967, A. Sakharov showed that to generate such an asymmetry, any model requires several necessary conditions [182]: (i) baryon number violation; (ii) C-symmetry and CP-symmetry violation; (iii) interactions out of thermal equilibrium. Each of these conditions can arise in the Standard Model.

Baryon number symmetry is anomalous as a consequence of instantons [183]. At zero temperature, the amplitude of the baryon-number-violating processes varies as $e^{-\frac{8\pi}{e^2}}$, which is too small to have any observable effect. At high temperatures such as in the early Universe, these transitions become unsuppressed [184]. The weak interactions of the SM violate C maximally and violate CP via the Kobayashi–Maskawa mixing. The CP violation, when appropriately normalized, is of order 10^{-20} and there are practically no kinematic enhancement factors in the thermal bath. Thus it is impossible to generate the required baryon asymmetry with such a small amount of CP violation. Consequently, baryogenesis implies that there must exist new sources of CP violation beyond the Kobayashi–Maskawa phase of the Standard Model. Within the Standard Model, departure from thermal equilibrium occurs at the electroweak phase transition [185]. However, this transition is not strongly first order, as required for successful baryogenesis. Thus, a different kind of departure from thermal equilibrium is required, which could result either from new physics or, alternatively, via a modification to the electroweak phase transition.

While the baryon number symmetry is anomalous in the SM, the difference of baryon number and lepton number (B-L) remains unbroken. Leptogenesis [186] is a class of scenarios where the baryon asymmetry of the Universe is produced from a lepton asymmetry generated in the decays of CP-violating interactions of the lightest of the heavy Majorana neutrinos, the seesaw partners of the ordinary neutrinos. An added bonus of this mechanism is that it ties the baryon asymmetry with neutrino properties. The rate of these CP-violating Yukawa interactions can be slow enough (that is, slower than the Hubble parameter, the expansion rate of the Universe, at the time that the asymmetry is generated) that departure from thermal equilibrium occurs. Lepton number violation comes from the Majorana masses of these new particles. For reviews, see [187, 188].

• Vacuum stability bound: The Standard Model scalar sector potential is rendered unstable if the 4-scalar coupling goes negative. The running of this coupling depends very sensitively on M_W, m_t and M_H. Using the currently measured values of m_t, M_W and M_H, the electroweak vacuum of the Standard Model becomes metastable at energy scales of the order of $\sim 10^{11}$ GeV decaying into a lower-energy vacuum at a higher energy with a lifetime larger than the age of the Universe [189, 190]. Appropriate modifications of the SM can stabilize the electroweak vacuum. This can be achieved in a wide variety of ways ranging from embedding the model in a more fundamental theory at a higher scale to simply adding a new heavy scalar singlet that acquires a large vacuum expectation value and has a quartic interaction with the ordinary Higgs doublet [191].

• Model for inflation: Cosmological inflation, or more simply inflation, is a theoretical framework that proposes an exponential expansion of space-time in the very early Universe [192, 193]. The inflationary epoch is posited to have lasted for 10^3–10^4 seconds. The slow roll inflationary paradigm (see, for example, [194] for a review) accounts for many of the observed properties of the Universe. It explains the large-scale isotropy, homogeneity and flatness of the Universe as well as accounting for the absence of any observed magnetic monopole. Furthermore, it provides a mechanism for the origin of the large-scale structure of the cosmos. Quantum fluctuations in a microscopic region get magnified to a cosmic size and become the seeds for the growth of structure in the Universe. Following the inflationary period, the Universe continued to expand, but at a slower rate. The new regions that come into view during this normal expansion phase are exactly the same regions that were pushed out of the horizon during inflation, and so they are at nearly the same temperature and curvature, because they come from the same originally small patch of space-time. Thus inflation explains why the temperatures and curvatures of different regions are so nearly equal. It also predicts that the total curvature of a space-slice at constant global time is zero. This prediction implies that the total ordinary matter, dark matter and residual vacuum energy in the Universe have to add up to the critical density, and the evidence supports this. Even more striking is that inflation allows calculation of the minute differences in temperature of different regions from quantum fluctuations during the inflationary era, and many of these quantitative predictions have been confirmed [195].

However, the particle physics model responsible for slow roll inflation is still lacking. Most models invoke a novel scalar degree of freedom, the inflaton, whose potential must be very flat to accommodate the observed size of said fluctuations (for a review, see [199]). In these models, this flatness is achieved by introducing extremely small couplings. Alternatively, attempts have been made to identify this scalar with the Standard Model Higgs boson. Such models require [200, 201] an extremely large nonminimal coupling of the Higgs bilinear to the Ricci scalar. Furthermore, the energy range in which these models represent a valid effective field theory is bounded above by a cutoff scale, which is found to be higher than the relevant dynamical scales throughout the whole history of the Universe, including the inflationary epoch and reheating [202]. Thus the extrapolation of the pure SM potential up to this scale is unwarranted and the scenario is analogous to other models of inflaton potentials afflicted with significant fine-tuning [203].

• Quantum theory of gravity: The minimal Standard Model ignores all gravitational interactions and treats space-time as flat. Nontrivial gravitational interactions can be introduced as a classical background in the Einstein–Hilbert action [3, 4, 204]

$$S_{EH} = \int d^4x \sqrt{-g} \, (4\pi G_N R + \mathcal{L}_M), \qquad (14.5)$$

where g is the determinant of the background metric tensor, R is the Ricci scalar and $\mathcal{L}_M$ contains the matter fields. Here $G_N = 6.67 \times 10^{-39}$ GeV^{-2} is the Newton gravitational constant. A deeper understanding of the origin of the Newton gravitational constant is at present lacking. In addition, there are other invariant terms containing additional space-time derivatives as well as a cosmological constant term. Using this action, amplitudes for processes with relevant energy scale E mediated by gravity can be expanded in a power series with an effective coupling strength of order $G_N E^2$. Thus at all presently accessible energies, the gravitational effects are very highly suppressed and can be safely ignored. However, as the Planck mass scale $M_{Pl} \sim \frac{1}{\sqrt{G_N}} \sim 10^{19}$ GeV is approached, this effective description breaks down as an infinite number of operators will all contribute nontrivially. To go beyond the classical background field approximation requires a quantum theory of gravity. The quest to secure a consistent quantum theory of gravity has plagued physicists since the advent of the quantum theory.

The quantization approach, which proved so successful in the development of QED and later of the Standard Model, which for the case of gravity requires adding fields with massless spin-2 graviton excitations coupled to the Standard Model, leads to internal inconsistencies [205–208] in conventional perturbation theory. Thus the current view is that general relativity and quantum mechanics are incompatible and, as yet undiscovered, further modifications are required. (An interesting effective field theory perspective is given in [209].) While there have been numerous attempts to find such extensions, none has yet proven satisfactory. For reviews of the theoretical landscape, see [210, 211]. On the other hand, techniques developed in some of these extensions have proven extremely useful in the allowing for the efficient computation of perturbative scattering amplitudes in gauge theory, in particular tree and one-loop multileg amplitudes in QCD. For reviews, see [212, 213].

• Outlook: At present, we are at a somewhat awkward time for theoretical particle physics. Encoded within the Standard Model, our current level of understanding has reached an unprecedented level. Many of the model consequences can be successfully extracted using perturbation theory. In some cases, such as QCD at low energies, a nonperturbative analysis is required. While lattice techniques have proven most useful for this vector gauge theory where the left- and right-handed quarks couple the same way to the gauge vectors, their utility is still somewhat limited when applied to chiral gauge theories such as the electroweak model, which have different couplings for the left- and right-handed fermions. As such, a nonperturbative definition of the SM is still lacking. Moreover, a nonperturbative understanding of strongly coupled chiral gauge theories could open potentially rich windows on possible extensions of the SM.

While the LHC continues to function excellently, providing ever more stringent tests of the Standard Model, nature has proven somewhat unkind in that the LHC has not revealed any novel particle states nor significant deviations from the SM. As such, much of the theoretical attention has shifted focus to questions that are very often far removed from possible experimental scrutiny.

These include structural issues regarding the nature of space-time, the origin of the gauge interactions accounting for the fundamental dynamics and properties in extreme environments inaccessible to experimental exploration as well as inquiries into multiverse theories and environmental selection. Ideally, any new physics resulting from such speculations should address the SM imperfections by predicting non–Standard Model outcomes of new proposed experiments while being consistent with existing data. Unfortunately, far too many of the proposals fail to meet this benchmark. On the other hand, there remain a variety of phenomena unanswered within the SM discussed in this Element, which (if nature allows) could be addressed using high-energy colliders, low-energy precision tests, observational cosmology, cosmic rays, dark-matter searches, gravitational waves, terrestrial and cosmic neutrinos and so on, which also merit further theoretical attention. In such enterprises, many currently diverse subfields of physics will need to share theoretical insights and experimental techniques. I am confident that symmetry will continue to play an instrumental role in any future advances.

Appendix

A.1 Notations and Conventions

We work in a system of units in which $\hbar = c = 1$. Thus energy, mass, inverse length and inverse time all carry the same dimension with conversion factors $1\,\mathrm{GeV} \simeq 2 \times 10^{-14}\,\mathrm{cm}^{-1} \simeq 6.57 \times 10^{-25}\,\mathrm{sec}^{-1} \simeq 1.45 \times 10^{18}\,\mathrm{gram}$.

We will consistently employ a summation convention where repeated indices are summed over.

A space-time point is denoted by

$$x^\mu = (t, \vec{r}) \tag{A.1}$$

so that the space-time derivative is

$$\partial_\mu \equiv \frac{\partial}{\partial x^\mu} = (\frac{\partial}{\partial t}, \nabla). \tag{A.2}$$

Greek indices $\mu, \nu, \ldots = 0, 1, 2, 3$ label the space-time dimensions, while Latin indices $i, j, \ldots = 1, 2, 3$ label only the spatial dimensions.

The (covariant) contravariant metric tensors $(\eta_{\mu\nu})\, \eta^{\mu\nu}$ are used to (lower) raise vector indices and are given by

$$\eta^{\mu\nu} = \begin{bmatrix} -1 & 0 & 0 & 0 \\ 0 & 1 & 0 & 0 \\ 0 & 0 & 1 & 0 \\ 0 & 0 & 0 & 1 \end{bmatrix}, \tag{A.3}$$

with $\eta_{\mu\nu}$ having the identical form.

The free Dirac equation and its conjugate are given by

$$(\gamma^\mu \frac{1}{i} \partial_\mu + m)\psi(x) = 0,$$

$$\overline{\psi}(x)(-\gamma^\mu \frac{1}{i} \overleftarrow{\partial}_\mu + m) = 0, \tag{A.4}$$

where

$$\overline{\psi}(x) = \psi^+(x)\gamma^0. \tag{A.5}$$

The four 4×4 γ^μ matrices satisfy

$$\{\gamma^\mu, \gamma^\nu\} = -2\eta^{\mu\nu}, \tag{A.6}$$

with

$$\gamma_0 \dagger = \gamma_0 \ , \ \gamma_i^\dagger = -\gamma_i,$$
$$\gamma_0^2 = 1 \ , \ \gamma_1^2 = \gamma_2^2 = \gamma_3^2 = -1. \tag{A.7}$$

It proves convenient to also define the matrix

$$\gamma_5 = i\gamma^0\gamma^1\gamma^2\gamma^3 = -\frac{i}{4!}\epsilon^{\mu\nu\lambda\rho}\gamma_\mu\gamma_\nu\gamma_\lambda\gamma_\rho, \tag{A.8}$$

which has the properties

$$\{\gamma_5, \gamma^\mu\} = 0,$$
$$\gamma_5^\dagger = \gamma_5,$$
$$\gamma_5^2 = 1. \tag{A.9}$$

Here $\epsilon^{\mu\nu\lambda\rho}$ is the Levi–Civita tensor density defined to be antisymmetric under the interchange of any two indices with $\epsilon^{0123} = 1$. Using the γ_5, the left-handed and right-handed projections of any Dirac field are

$$\psi_L(x) = \frac{1}{2}(1 - \gamma_5)\psi(x),$$
$$\psi_R(x) = \frac{1}{2}(1 + \gamma_5)\psi(x). \tag{A.10}$$

Finally, we define the matrices

$$\sigma_{\mu\nu} = \frac{i}{2}[\gamma_\mu, \gamma_\nu] = -\sigma_{\nu\mu}. \tag{A.11}$$

The 16 matrices

$$\Gamma = \{1, \gamma_5, \gamma^\mu, \gamma^\mu\gamma_5, \sigma^{\mu\nu}\} \tag{A.12}$$

form a linearly independent set of 4×4 matrices.

The solution to the free Dirac equation has the spinor decomposition

$$\psi(x) = \sum_{\pm s} \int \frac{d^3p}{(2\pi)^3 2p^0} \left(e^{ipx}b(p,s)U(p,s) + e^{ipx}d^\dagger(p,s)V(p,s)\right), \tag{A.13}$$

where $px = -p^0x^0 + \vec{p} \cdot \vec{x}$ with $p^0 = \sqrt{\vec{p}^2 + m^2}$.

The four-component spinors $U(p,s)$ and $V(p,s)$ obey

$$(\gamma p + m)U(p,s) = 0 = \overline{U}(p,s)(\gamma p + m), \quad \overline{U}(p,s) = U^\dagger(p,s)\gamma^0, \tag{A.14}$$
$$(-\gamma p + m)V(p,s) = 0 = \overline{V}(p,s)(-\gamma p + m), \quad \overline{V}(p,s) = V^\dagger(p,s)\gamma^0, \tag{A.15}$$

and

$$-\gamma_5 \gamma s U(p,s) = U(p,s), \tag{A.16}$$
$$-\gamma_5 \gamma s V(p,s) = V(p,s). \tag{A.17}$$

It follows that the projectors

$$\Sigma(s) = \frac{1 - \gamma_5\gamma s}{2} \tag{A.18}$$

satisfy the covariant spin eigenvalue equations

$$\Sigma(s)U(p,s) = U(p,s),$$
$$\Sigma(s)U(p,-s) = 0,$$
$$\Sigma(s)V(p,s) = V(p,s),$$
$$\Sigma(s)V(p,-s) = 0, \tag{A.19}$$

and consequently projects out the single helicity component $+s$. The polarization vector s^μ and the momentum vector p^μ are obtained from their rest frame values $\hat{s}^\mu = (0, \hat{s})$, $\hat{p}^\mu = (m, 0)$ via the Lorentz transformation. The unit vector $\hat{s}$ points in the direction of the spin of the particle. It follows that, in any Lorentz frame,

$$s_\mu s^\mu = 1,$$
$$s_\mu p^\mu = 0,$$
$$p_\mu p^\mu = -m^2. \tag{A.20}$$

The spinors also satisfy the orthogonality relations

$$\overline{U}(p,s)U(p,s') = 2m\delta_{ss'},$$
$$\overline{V}(p,s)V(p,s') = -2m\delta_{ss'},$$
$$\overline{U}(p,s)V(p,s') = 0, \tag{A.21}$$

and the completeness relations

$$\frac{1}{2m} \sum_{\pm s} U(p,s)\overline{U}(p,s) = \frac{-\gamma p + m}{2m},$$
$$\frac{1}{2m} \sum_{\pm s} V(p,s)\overline{V}(p,s) = \frac{-\gamma p - m}{2m}. \tag{A.22}$$

A.2 The $SU(2)_L \times U(1)$ Electroweak Model

The specification of the $SU(2)_L \times U(1)$ transformation properties completely fixes the possible renormalizable interactions. These can be classified into five categories:

a): Gauge Sector

b): Scalar Sector

c): Gauge-Scalar Sector

d): Gauge-Fermion Sector

e): Fermion-Scalar Sector

We now consider each in turn.

a) Yang–Mills Lagrangian and Gauge Boson Self-Interactions

The $SU(2)_L \times U(1)$ Yang–Mills theory with gauge couplings g_2 and g_1 respectively is constructed using the $SU(2)_L$ gauge fields W_a^μ, $a = 1, 2, 3$, and the $U(1)$ gauge field Y^μ. Under the local $SU(2)_L \times U(1)$ gauge group, these fields, besides rotating appropriately, also have inhomogeneous pieces that allow for the establishment of the local symmetries. Specifically, their infinitesimal transformations take the form

$$W_a^\mu \to W_a^\mu - g_2 \epsilon_{abc} \delta\omega_b W_c^\mu + \partial^\mu \delta\omega_a,$$
$$Y^\mu \to Y^\mu + \partial^\mu \delta y. \tag{A.1}$$

Here ϵ_{abc} is the Levi–Civita tensor, while $\delta\omega_a$, $a = 1, 2, 3$ and δy are the infinitesimal parameters associated with the $SU(2)_L$ and $U(1)$ groups. The locally invariant $SU(2)_L \times U(1)$ Yang–Mills Lagrangian is then constructed in terms of the gauge covariant field strengths

$$W_a^{\mu\nu} = \partial^\mu W_a^\nu - \partial^\nu W_a^\mu + g_2 \epsilon_{abc} W_b^\mu W_c^\nu,$$
$$Y^{\mu\nu} = \partial^\mu Y^\nu - \partial^\nu Y^\mu, \tag{A.2}$$

as

$$\mathcal{L}_{YM} = -\frac{1}{4} W_a^{\mu\nu} W_{a\mu\nu} - \frac{1}{4} Y^{\mu\nu} Y_{\mu\nu}. \tag{A.3}$$

Note that to guarantee that $W_a^{\mu\nu}$ transforms as an $SU(2)_L$ vector requires the inclusion of terms quadratic in W_a^μ. This in turn leads to self-interactions among the gauge fields in $\mathcal{L}_{YM}$. This is a characteristic feature of non-Abelian gauge theories.

b) Scalar Sector and Spontaneous Symmetry Breaking

In the Standard Model, the spontaneous $SU(2)_L \times U(1)$ symmetry breakdown is engendered by introducing a complex $SU(2)$ doublet of scalar fields,

$$\Phi = \begin{bmatrix} \phi_0 \\ \phi_- \end{bmatrix}. \tag{A.4}$$

Here the subscripts denote the electric charges of the components of the doublet. Through its self-interactions, Φ acquires a nonvanishing vacuum expectation value for the neutral component so that

$$\langle \Phi \rangle = \frac{v}{\sqrt{2}} \begin{bmatrix} 1 \\ 0 \end{bmatrix}, \tag{A.5}$$

thus spontaneously breaking the symmetry.

To achieve this vacuum structure, it suffices to choose an $SU(2)_L \times U(1)$ invariant potential, $V(\Phi)$, of the form

$$V(\Phi) = \lambda\left(\Phi^\dagger\Phi - \frac{v^2}{2}\right)^2.$$

(A.6)

Since $T_a\langle\Phi\rangle \neq 0$ and $Y\langle\Phi\rangle \neq 0$, the $SU(2)_L \times U(1)$ symmetry is spontaneously broken. However, because only the neutral component of Φ develops the vacuum value, it follows that

$$Q_{EM}\langle\Phi\rangle = (T_3 + Y)\langle\Phi\rangle = \begin{bmatrix} 0 & 0 \\ 0 & -1 \end{bmatrix} \frac{v}{\sqrt{2}} \begin{bmatrix} 1 \\ 0 \end{bmatrix} = 0$$

(A.7)

and electromagnetism remains unbroken.

c) Gauge-Scalar Sector and Vector Mass Generation

The electric charge generator (see Eq. (9.10))

$$Q_{EM} = T_3 + Y$$

(A.8)

dictates the infinitesimal $U(1)$ transformation properties assignments of the scalar fields so as to produce the electric charges of the various components as indicated by the subscripts in Eq. (A.4) so that

$$\Phi \rightarrow \left(1 + ig_2\frac{\sigma_a}{2}\delta w_a - ig_1\frac{1}{2}\delta y\right)\Phi,$$

(A.9)

where σ_a are the Pauli matrices. Defining the Φ covariant derivative as

$$D_\mu\Phi = \left(\partial_\mu - ig_2\frac{\sigma_a}{2}W_{a_\mu} + ig_1\frac{1}{2}Y_\mu\right)\Phi,$$

(A.10)

a locally $SU(2)_L \times U(1)$ invariant Lagrangian is constructed as

$$\mathcal{L}_{G-H} = -(D_\mu\Phi)^\dagger(D^\mu\Phi) - V(\Phi),$$

(A.11)

where $V(\Phi)$ is the scalar potential of Eq. (A.6).

The $SU(2)_L \times U(1) \rightarrow U(1)_{em}$ spontaneous symmetry breakdown results in three of the gauge fields acquiring mass via the Higgs mechanism. This is readily established by replacing Φ by its vacuum value $\langle\Phi\rangle$ in Eq. (A.11) giving

$$\mathcal{L}_{\substack{vector \\ mass}} = -\left(D_\mu\langle\Phi\rangle\right)^\dagger\left(D^\mu\langle\Phi\rangle\right)$$

$$= -\langle\Phi^\dagger\rangle\left(g_2\frac{\sigma_a}{2}W_a^\mu - g_1\frac{1}{2}Y^\mu\right)\left(g_2\frac{\sigma_b}{2}W_{b_\mu} - g_1\frac{1}{2}Y_\mu\right)\langle\Phi\rangle.$$

(A.12)

It proves convenient to define the complex fields

$$W_\pm^\mu = \frac{1}{\sqrt{2}}(W_1^\mu \pm iW_2^\mu)$$

(A.13)

and the real orthogonal combinations

$$Z^\mu = \cos\theta_W W_3^\mu - \sin\theta_W Y^\mu,$$
$$A^\mu = \sin\theta_W W_3^\mu + \cos\theta_W Y^\mu, \tag{A.14}$$

where the weak mixing angle θ_W, the Weinberg angle, satisfies

$$\tan\theta_W = \frac{g_1}{g_2}. \tag{A.15}$$

These transformations diagonalize the vector mass term, giving

$$\mathcal{L}_{\substack{\text{vector}\\\text{mass}}} = -M_W^2 W_+^\mu W_{-\mu} - \frac{1}{2}M_Z^2 Z^\mu Z_\mu, \tag{A.16}$$

with

$$M_W = \frac{g_2 v}{2} \quad ; \quad M_Z = \frac{g_2 v}{2\cos\theta_W}. \tag{A.17}$$

Thus, the $W_\pm$ and Z fields describe massive vector modes. Note that the vector masses satisfy the simple relation

$$M_W = M_Z \cos\theta_W, \tag{A.18}$$

which is a consequence of implementing the symmetry breaking using a doublet of scalar fields. Since the A^μ fields has no mass term, it is natural to identify it with the photon field.

The gauge boson self-interactions arising in the Yang–Mills Lagrangian Eq. (A.3) can be written in terms of the physical $W_\pm^\mu$, Z^μ and A^μ degrees of freedom. It is convenient to split these interactions into terms that are trilinear and quadrilinear in the gauge fields. One finds

$$\mathcal{L}_{\text{trilinear}} = ie\left\{\eta_{\mu\nu}\left[-\partial_\lambda(A^\mu + \cot\theta_W Z^\mu)(W_+^\nu W_-^\lambda - W_+^\lambda W_-^\nu)\right.\right.$$
$$+ (A^\mu + \cot\theta_W Z^\mu)(\partial_\lambda W_+^\nu W_-^\lambda - W_+^\lambda \partial_\lambda W_-^\nu)\Big]$$
$$+ \eta_{\nu\lambda}(A^\mu + \cot\theta_W Z^\mu)(W_+^\nu \partial_\mu W_-^\lambda - \partial_\mu W_+^\nu W_-^\lambda)\Big\}, \tag{A.19}$$

$$\mathcal{L}_{\text{quadrilinear}} = \frac{e^2}{4\sin^2\theta_W}\left\{(2\eta_{\mu\lambda}\eta_{\nu\rho} - \eta_{\mu\rho}\eta_{\nu\lambda} - \eta_{\mu\nu}\eta_{\lambda\rho})W_+^\mu W_-^\nu W_+^\lambda W_-^\rho\right.$$
$$- 2\sin^2\theta_W(2\eta_{\mu\nu}\eta_{\lambda\rho} - \eta_{\mu\lambda}\eta_{\nu\rho} - \eta_{\mu\rho}\eta_{\nu\lambda})$$
$$\times (A^\mu + \cot\theta_W Z^\mu)(A^\nu + \cot\theta_W Z^\nu)W_+^\lambda W_-^\rho\Big\}. \tag{A.20}$$

The complex doublet field Φ is composed of four real fields. Of these, three are absorbed into the longitudinal components of the $W_\pm$ and Z fields. The remaining real scalar field describes the physical Higgs boson H. To make the physical content manifest, the Φ doublet can be parameterized in terms of the real fields $\vec{\xi}$ and H as

$$\Phi = \exp\left(i\frac{\vec{\sigma}\cdot\vec{\xi}}{v}\right)\frac{1}{\sqrt{2}}\begin{bmatrix} v+H \\ 0 \end{bmatrix}. \tag{A.21}$$

The $\vec{\xi}$ fields, however, are unphysical and can be completely eliminated by making an appropriate gauge choice referred to as the unitary gauge. This follows since $\exp(i\frac{\vec{\sigma}\cdot\vec{\xi}}{2})$ is an $SU(2)_L$ group element. Thus, after making the finite $SU(2)_L$ gauge transformation

$$\Phi \to e^{-i\frac{\vec{\sigma}\cdot\vec{\xi}}{v}}\Phi = \frac{1}{\sqrt{2}}\begin{bmatrix} v+H \\ 0 \end{bmatrix}, \tag{A.22}$$

the $\vec{\xi}$ fields have completely disappeared. Substituting this unitary gauge representation for Φ into Eq. (A.11) gives the Higgs scalar kinetic term, the vector mass term and the Higgs-vector interaction term

$$\mathcal{L}_{HV} = -g_2 M_W W_+^\mu W_{-\mu} H - \frac{g_2 M_Z}{2\cos\theta_W} Z^\mu Z_\mu H$$
$$- \frac{g_2^2}{4} W_+^\mu W_{-\mu} H^2 - \frac{g_2^2}{8\cos^2\theta_W} Z^\mu Z_\mu H^2. \tag{A.23}$$

Note that there is no direct coupling of the Higgs scalar H to the photon field A^μ. Finally, the scalar potential, Eq. (A.6), reduces to

$$V(H) = \lambda v^2 H^2 + \lambda v H^3 + \frac{\lambda}{4} H^4. \tag{A.24}$$

While the unitary gauge introduced here directly allows one to glean the physical degrees of freedom, it is not very useful for performing higher radiative (loop) corrections. For one thing, the massive vector propagator is of the Proca form, which does not exhibit falloff at large momenta. However, since we are dealing with a gauge theory, other gauge choices can be employed which do have better falloff. A particularly useful class for performing higher-order calculations and proving the renormalizability of the theory is the R_ξ gauge. Here one parametrizes the Higgs doublet as

$$\Phi = \frac{1}{\sqrt{2}}\begin{bmatrix} v + \chi_1 + i\chi_2 \\ \chi_3 + i\chi_4 \end{bmatrix} \equiv \begin{bmatrix} \frac{1}{\sqrt{2}}v + s_0 \\ s_- \end{bmatrix}, \tag{A.25}$$

with the χ_i Hermitian fields. One next adds the gauge-fixing Lagrangian

$$\mathcal{L}_{\text{gauge fixing}} = \frac{\xi_A}{2}\left(\partial_\mu A^\mu\right)^2 + \frac{\xi_Z}{2}\left(\partial_\mu Z^\mu + \frac{M_Z}{\xi_Z}\chi_2\right)^2$$
$$+ \xi_W\left(\partial_\mu W_+^\mu + \frac{iM_W}{\xi_W}s_+\right)\left(\partial_\nu W_-^\nu - \frac{iM_W}{\xi_W}s_-\right). \tag{A.26}$$

When used in the path integral formulation of the QFT [5], the addition of this gauge-fixing term also gives rise to a functional Jacobian, which must be correctly taken into account. This functional determinant can be cast into a local Lagrangian contribution by introducing the fictitious anticommuting scalar Faddeev–Popov ghost fields [214]. Thus the R_ξ gauge fixed electroweak Lagrangian contains many unphysical degrees of freedom. One can show that all these unphysical degrees of freedom cancel when calculating S-matrix elements of the physical degrees of freedom. For any $\xi \neq 0$, all the field propagators have a $1/k^2$ falloff for large k^2. For example, the vector propagators in R_ξ take the form

$$\Delta_{\mu\nu}^{(W)}(k) = \frac{1}{k^2 + M_W^2 - i\epsilon} \left(\eta_{\mu\nu} + \frac{k_\mu k_\nu (1 + \xi_W)}{M_W^2 - k^2 \xi_W} \right),$$

$$\Delta_{\mu\nu}^{(Z)}(k) = \frac{1}{k^2 + M_Z^2 - i\epsilon} \left(\eta_{\mu\nu} + \frac{k_\mu k_\nu (1 + \xi_Z)}{M_Z^2 - k^2 \xi_Z} \right),$$

$$\Delta_{\mu\nu}^{(A)}(k) = \frac{1}{k^2 - i\epsilon} \left(\eta_{\mu\nu} - \frac{k_\mu k_\nu (1 + \xi_A)}{k^2 \xi_A} \right). \tag{A.27}$$

Unitary gauge formally corresponds to the limit $\xi \to 0$.

The R_ξ gauge-fixing term and ghost Lagrangian clearly break the gauge invariance of the underlying gauge-invariant action. However, they do so in just such a way that the S-matrix elements of the physical degrees of freedom are gauge invariant. In fact, while the gauge invariance is broken, there remains an unbroken global symmetry of the effective action comprised of the original gauge-invariant action, the R_ξ gauge-fixing term and ghost Lagrangian. This is the BRST symmetry [215, 216]. The BRST transformations of the gauge fields, scalars and fermions take the same form as their gauge transformations, except the gauge parameters are replaced by the product of the anticommuting ghost fields and space-time–independent anticommuting (Grassmann) parameters of the BRST transformation. In addition, the gauge-invariant ghost fields themselves have nontrivial BRST transformation properties so constructed as to render the effective action BRST invariant. Since the Grassmann parameters are space-time independent, the BRST transformation is global. But it is not a conventional global symmetry, due to the Grassmann nature of the parameters. This BRST global symmetry is the underlying symmetry of the action comprised of the sum of the orginal gauge-invariant action and the gauge-fixing and ghost terms and can be used to show that all the unphysical degrees of freedom cancel in the physical S-matrix elements.

d) Gauge-Fermion Sector and the Low-Energy Effective Lagrangian

The infinitesimal $SU(2)_L \times U(1)$ transformations of the fermions follow immediately from their quantum numbers, which are given in Tables 9.1 and 9.2:

$$L_{iL} = \begin{bmatrix} v_{iL} \\ e_{iL} \end{bmatrix} \rightarrow \left(1 + ig_2\frac{\sigma_a}{2}\delta w_a - ig_1\frac{1}{2}\delta y\right) L_{iL}$$

$$e_{iR} \rightarrow (1 - ig_1\delta y)\, e_{iR}$$

$$Q_{iL} = \begin{bmatrix} u_{iL} \\ d_{iL} \end{bmatrix} \rightarrow \left(1 + ig_2\frac{\sigma_a}{2}\delta w_a - ig_1\frac{1}{6}\delta y\right) Q_{iL}$$

$$u_{iR} \rightarrow \left(1 + ig_1\frac{2}{3}\delta y\right) u_{iR}$$

$$d_{iR} \rightarrow \left(1 - ig_1\frac{1}{3}\delta y\right) d_{iR}. \tag{A.28}$$

Here the subscript $i = 1, 2, 3$ labels the generation. Defining the covariant derivatives for the lepton and quark fields as

$$D_\mu L_{iL} = \left(\partial_\mu - ig_2\frac{\sigma_a}{2}W_{a\mu} + ig_1\frac{1}{2}Y_\mu\right) L_{iL},$$

$$D_\mu e_{iR} = \left(\partial_\mu + ig_1 Y_\mu\right) e_{iR},$$

$$D_\mu Q_{iL} = \left(\partial_\mu - ig_2\frac{\sigma_a}{2}W_{a\mu} - ig_1\frac{1}{6}Y_\mu\right) Q_{iL},$$

$$D_\mu u_{iR} = \left(\partial_\mu - ig_1\frac{2}{3}Y_\mu\right) u_{iR},$$

$$D_\mu d_{iR} = \left(\partial_\mu + ig_1\frac{1}{3}Y_\mu\right) d_{iR}, \tag{A.29}$$

a locally $SU(2)_L \times U(1)$ invariant Lagrangian which generalizes the fermionic kinetic terms is simply obtained as

$$\mathcal{L}_{F-G} = \sum_i \left(-\bar{L}_{iL}\gamma^\mu\frac{1}{i}D_\mu L_{iL} - \bar{e}_{iR}\gamma^\mu\frac{1}{i}D_\mu e_{iR} \right.$$
$$\left. - \bar{Q}_{iL}\gamma^\mu\frac{1}{i}D_\mu Q_{iL} - \bar{u}_{iR}\gamma^\mu\frac{1}{i}D_\mu u_{iR} - \bar{d}_{iR}\gamma^\mu\frac{1}{i}D_\mu d_{iR} \right). \tag{A.30}$$

The fermion coupling to the gauge fields is thus dictated by the local $SU(2)_L \times U(1)$ invariance to be

$$\mathcal{L}_{int} = g_2 W_a^\mu J_{a\mu} + g_1 Y^\mu J_{Y_\mu}, \tag{A.31}$$

where the $SU(2)_L$ and $U(1)$ fermionic currents are

$$J_a^\mu = \sum_i \left(\bar{L}_{iL}\gamma^\mu\frac{\sigma_a}{2}L_{iL} + \bar{Q}_{iL}\gamma^\mu\frac{\sigma_a}{2}Q_{iL} \right), \tag{A.32}$$

$$J_Y^\mu = \sum_i \left(-\frac{1}{2}\bar{L}_{iL}\gamma^\mu L_{iL} - \bar{e}_{iR}\gamma^\mu e_{iR} \right.$$
$$\left. + \frac{1}{6}\bar{Q}_{iL}\gamma^\mu Q_{iL} + \frac{2}{3}\bar{u}_{iR}\gamma^\mu u_{iR} - \frac{1}{3}\bar{d}_{iR}\gamma^\mu d_{iR} \right). \tag{A.33}$$

Using Eqs. (A.13)–(A.15), $\mathcal{L}_{\text{int}}$ can be rewritten as

$$\mathcal{L}_{\text{int}} = eJ^{\mu}_{EM}A_{\mu} + \frac{e}{2\sqrt{2}\sin\theta_W}(W^{\mu}_{+}J_{-\mu} + W^{\mu}_{-}J_{+\mu})$$

$$+ \frac{e}{2\cos\theta_W\sin\theta_W}Z_{\mu}J^{\mu}_{NC}. \tag{A.34}$$

Here we have identified the electric charge

$$e = g_1\cos\theta_W = g_2\sin\theta_W \tag{A.35}$$

as the coupling of the photon to the electromagnetic current

$$J^{\mu}_{EM} = J^{\mu}_3 + J^{\mu}_Y$$

$$= \sum_i \left(-\overline{e}_i\gamma^{\mu}e_i + \frac{2}{3}\overline{u}_i\gamma^{\mu}u_i - \frac{1}{3}\overline{d}_i\gamma^{\mu}d_i\right), \tag{A.36}$$

whose form is consistent with the electric charge definition $Q_{EM} = T_3 + Y$. The weak charged currents are

$$J^{\mu}_{\pm} = 2(J^{\mu}_1 \pm iJ^{\mu}_2) \tag{A.37}$$

so that

$$J^{\mu}_{+} = \sum_i \left(\overline{e_{iL}}\gamma^{\mu}\nu_{iL} + \overline{d_{iL}}\gamma^{\mu}u_{iL}\right),$$

$$J^{\mu}_{-} = \sum_i \left(\overline{\nu_{iL}}\gamma^{\mu}e_{iL} + \overline{d_{iL}}\gamma^{\mu}u_{iL}\right), \tag{A.38}$$

while the weak neutral currents

$$J^{\mu}_{NC} = 2(J^{\mu}_3 - \sin^2\theta_W J^{\mu}_{em})$$

$$= \sum_i \Bigg(\overline{\nu_{iL}}\gamma^{\mu}\nu_{iL} - (1 - 2\sin^2\theta_W)\overline{e_{iL}}\gamma^{\mu}e_{iL} + 2\sin^2\theta_W\overline{e_{iR}}\gamma^{\mu}e_{iR}$$

$$+ (1 - \frac{4}{3}\sin^2\theta_W)\overline{u_{iL}}\gamma^{\mu}u_{iL} - \frac{4}{3}\sin^2\theta_W\overline{u_{iR}}\gamma^{\mu}u_{iR}$$

$$- (1 - \frac{2}{3}\sin^2\theta_W)\overline{d_{iL}}\gamma^{\mu}d_{iL} + \frac{2}{3}\sin^2\theta_W\overline{d_{iR}}\gamma^{\mu}d_{iR}\Bigg) \tag{A.39}$$

have also been identified through their couplings to the vector bosons $W^{\mp}$ and Z respectively.

For weak interaction processes whose relevant momentum transfer q^{μ} is such that $q^2 \ll M^2_W, M^2_Z$, the effects of $\mathcal{L}_{\text{int}}$ can be well approximated using second-order perturbation theory, leading to the current-current effective Lagrangian

$$\mathcal{L}^{\text{weak}}_{\text{eff}} = \left(\frac{e}{2\sqrt{2}\sin\theta_W}\right)^2 \frac{1}{M^2_W}J^{\mu}_{+}J_{-\mu}$$

$$+ \frac{1}{2} \left(\frac{e}{2 \cos \theta_W \sin \theta_W} \right)^2 \frac{1}{M_Z^2} J_{NC}^\mu J_\mu^{NC}. \tag{A.40}$$

Thus in this limit, the Glashow–Salam–Weinberg model reduces to the Fermi theory plus an additional neutral current contribution. This allows the identification of the Fermi constant G_F as

$$\frac{G_F}{\sqrt{2}} = \frac{e^2}{8 \sin^2 \theta_W M_W^2}. \tag{A.41}$$

Thus Eq. (A.34) takes the form

$$\mathcal{L}_{\text{eff}}^{\text{weak}} = \frac{G_F}{\sqrt{2}} \left(J_+^\mu J_{-\mu} + J_{NC}^\mu J_\mu^{NC} \right). \tag{A.42}$$

e) Fermion-Scalar Sector, Fermion Mass Generation, and Quark Flavor Mixing

Besides giving mass to the $W_\pm$ and Z vector bosons, the same scalar doublet can also provide mass terms for the fermions. To appreciate this point, note that the (Dirac) fermion mass term connects right and left fields. Since the quark and lepton fields of opposite chirality transform differently under $SU(2)_L \times U(1)$, no such mass terms are permitted before the symmetry is spontaneously broken. However, since Φ is an $SU(2)_L$ doublet, it can have $SU(2)_L \times U(1)$ invariant couplings with the right- and left-handed fermion fields via Yukawa interactions.

It proves convenient to introduce the charge conjugate fields

$$\tilde{\Phi} = i\sigma_2 \Phi^* = \begin{bmatrix} \phi_+ \\ -\phi_0^* \end{bmatrix}. \tag{A.43}$$

It is easy to check that $\tilde{\Phi}$ transforms under an infinitesimal $SU(2)_L \times U(1)$ transformation as

$$\tilde{\Phi} \to \left(1 + i\delta\omega_a \frac{\sigma_a}{2} + i\delta y \frac{1}{2} \right) \tilde{\Phi}. \tag{A.44}$$

That is, it is also an $SU(2)_L$ doublet, but it has weak hypercharge equal to $+\frac{1}{2}$. The most general $SU(2)_L \times U(1)$ invariant Yukawa interaction is then constructed as

$$\begin{aligned}
\mathcal{L}_{\text{Yukawa}} = \sum_{i,j} \Big(&\Gamma_{ij}^e \overline{L}_{iL} \tilde{\Phi} e_{jR} + \Gamma_{ij}^{e*} \overline{e}_{jR} \tilde{\Phi}^\dagger L_{iL} \\
&- \Gamma_{ij}^u \overline{Q}_{iL} \Phi u_{jR} - \Gamma_{ij}^{u*} \overline{u}_{jR} \Phi^\dagger Q_{iL} \\
&+ \Gamma_{ij}^d \overline{Q}_{iL} \tilde{\Phi} d_{jR} + \Gamma_{ij}^{d*} \overline{d}_{jR} \tilde{\Phi}^\dagger Q_{iL} \Big),
\end{aligned} \tag{A.45}$$

where the Γ_{ij}^f are generalized Yukawa couplings. (Here $i,j = 1,2,3$ are generation labels.) Using the unitary gauge representations

$$\Phi = \frac{1}{\sqrt{2}} \begin{bmatrix} v + H \\ 0 \end{bmatrix} \quad ; \quad \tilde{\Phi} = -\frac{1}{\sqrt{2}} \begin{bmatrix} 0 \\ v + H \end{bmatrix}, \tag{A.46}$$

this yields mass terms for the fermions of the same charge, as well as couplings
of the Higgs boson H with the fermions. The fermion mass term reads

$$\mathcal{L}_{\substack{\text{fermion} \\ \text{mass}}} = \sum_{i,j} \left(-\bar{e}_{iL} M_{ij}^e e_{jR} - \bar{u}_{iL} M_{ij}^u u_{jR} - \bar{d}_{iL} M_{ij}^d d_{jR} \right) + h.c., \tag{A.47}$$

with

$$M_{ij}^f = \frac{1}{\sqrt{2}} \Gamma_{ij}^f v , \quad (f = e, u, d). \tag{A.48}$$

Since there is no right-handed neutrino included in the model, no neutrino mass
term ensues. Note that since the Yukawa couplings Γ_{ij}^f are arbitrary parameters,
so are the entries in the mass matrices M_{ij}^f. Because these mass matrices are not
in general diagonal, it is necessary to perform a basis change in order to obtain
the physical, mass diagonal, fermion fields. This basis change leads to a mixing
of states between different families, which is probed by the weak interactions.
This is the origin of the Cabbibo–Kobayashi–Maskawa (CKM) weak mixing
angles [93, 96] in the Standard Model.

The matrices M^f can be diagonalized by the bi-unitary transformation
$((U_L^f)^\dagger U_L^f = I; \ (U_R^f)^\dagger U_R^f = I)$ as

$$(U_L^f)^\dagger M^f U_R^f = (M^f)_{\text{diag}}, \quad (f = e, u, d), \tag{A.49}$$

which corresponds to the fermion field change of basis

$$\psi_{L,R}^f \rightarrow U_{L,R}^f \psi_{L,R}^f , \quad (f = e, u, d). \tag{A.50}$$

(The transformation is unitary if M^f is a Hermitian matrix.)

In addition to diagonalizing the fermion mass matrix, these replacements leave
the neutral and electromagnetic currents unaltered, but introduce a mixing
matrix for the charged currents. This is readily established. Focusing on the
three-generation case, the charged current J_-^μ before the basis change is

$$J_-^\mu = 2[\bar{v}_e, \bar{v}_\mu, \bar{v}_\tau]_L \gamma^\mu 1 \begin{bmatrix} e \\ \mu \\ \tau \end{bmatrix}_L + 2[\bar{u}, \bar{c}, \bar{t}]_L \gamma^\mu 1 \begin{bmatrix} d \\ s \\ b \end{bmatrix}_L. \tag{A.51}$$

After the basis change of Eq. (A.50), this current becomes

$$J^\mu_- = 2[\bar\nu_e, \bar\nu_\mu, \bar\nu_\tau]_L \gamma^\mu U^e_L \begin{bmatrix} e \\ \mu \\ \tau \end{bmatrix}_L + 2(\bar u, \bar c, \bar t)_L \gamma^\mu (U^u_L)^\dagger (U^d_L) \begin{bmatrix} d \\ s \\ b \end{bmatrix}_L . \quad (A.52)$$

Because the neutrinos are massless, the matrix U^e_L can be eliminated by a redefinition of the neutrino fields. On the other hand, the unitary matrix

$$V_{CKM} = (U^u_L)^\dagger U^d_L \quad (A.53)$$

appearing in the quark sector remains. It is the 3×3 unitary, $V^\dagger_{CKM} V_{CKM} = I$, mixing matrix of the charged weak current: the Cabbibo–Kobayashi–Maskawa mixing matrix. It is also clear that the basis change, Eq. (A.50), does not alter the form of the neutral or electromagnetic currents since they always connect fermions of the same charge and

$$(U^f_L)^\dagger U^f_L = (U^f_R)^\dagger U^f_R = 1. \quad (A.54)$$

Since diagonalizing the mass matrices M^f is equivalent to diagonalizing the Yukawa couplings Γ^f, it follows that after the basis change to the physical fermion fields, the coupling of the Higgs boson H to the fermions is purely diagonal and the Higgs-fermion Lagrangian is simply given by

$$\mathcal{L}_{HF} = -\sum_i (m_{e_i} \bar e_i e_i + m_{u_i} \bar u_i u_i + m_{d_i} \bar d_i d_i)\left(1 + \frac{H}{v}\right). \quad (A.55)$$

Note that the Higgs-fermion couplings are proportional to the mass of the fermion with which it interacts. Thus the only remnant of any nondiagonal Yukawa interaction appearing in Eq. (A.45) is through the unitary mixing matrix V entering in the charged weak current

$$J^\mu_- = 2 [\bar\nu_e, \bar\nu_\mu, \bar\nu_\tau] \gamma^\mu \begin{bmatrix} e \\ \mu \\ \tau \end{bmatrix}_L + 2 [\bar u, \bar c, \bar t]_L \gamma^\mu V_{CKM} \begin{bmatrix} d \\ s \\ b \end{bmatrix}_L . \quad (A.56)$$

The combination of a diagonal neutral current and charged weak current with unitary mixing automatically guarantees the GIM [91] mechanism. It is extremely pleasing to see that this mechanism, which was invented to enforce the experimentally mandated suppression of flavor-changing neutral current processes, emerges in so natural a fashion in the standard electroweak theory.

It is straightforward to count the number of independent parameters characterizing the Cabbibo–Kobayashi–Maskawa mixing matrix. In general, an $n \times n$ unitary matrix is specified by n^2 parameters. Of these, $\frac{1}{2}n(n-1)$ are real angles and $\frac{1}{2}n(n+1)$ are phases. For the n family case, $2n-1$ of these phases can be removed by diagonal phase redefinitions of the $2n$ quark fields. (Only $2n-1$ and not $2n$ phases can be eliminated, since an overall phase has no physical content.) Thus for n families, the Cabbibo–Kobayashi–Maskawa mixing matrix V is parameterized by $\frac{1}{2}n(n-1)$ real angles and $\frac{1}{2}(n-1)(n-2)$ phases. For two generations, there is only one angle involved, which is the Cabibbo angle. For the three-generation case, which describes the world as we now know it, this counting gives three real angles and a single phase. This phase allows for CP-symmetry-violating processes.

References

[1] Atlas Collaboration: G. Aad, T. Abajyan, B. Abbott et al., Phys. Lett. **B716** (2012) 1–29. e-Print: 1207.7214 [hep-ex].

[2] CMS Collaboration: S. Chatrchyan, V. Khachatryan, A. M. Sirunyan et al. Phys. Lett. **B716** (2012) 30–61. e-Print: 1207.7235 [hep-ex].

[3] A. Einstein, Preussische Akademie der Wissenschaften, *Sitzungs-berichte der Preussischen Akademie der Wissenschaften* (Berlin), Part 2 (in German) (1915) 844–847.

[4] A. Einstein, Annalen der Physik (in German) **354** (1916) 769.

[5] M. E. Peskin and D. V. Schroeder, *An Introduction to Quantum Field Theory* (Westview Press, Nashville, 1995).

[6] E. C. G. Stueckelberg and A. Petermann, Helv. Phys. Acta (in French) **26** (1953) 499–520.

[7] M. Gell-Mann and F. E. Low, Phys. Rev. **95** (1954) 1300–1312.

[8] K. Wilson, Phys. Rev. **179** (1969) 1499.

[9] K. Wilson and J. Kogut, Phys. Rept. **12** (1974) 75–200.

[10] E. Wigner, *Group Theory and Its Application to the Quantum Mechanics of Atomic Spectra* (Academic Press, Cambridge, MA, 1959).

[11] H. Georgi, *Lie Algebras in Particle Physics* (Benjamin/Cummings Publishing Company, San Francisco, 1982).

[12] E. Noether, Mathematisch-Physikalische Klasse (in German) (1918) 236–257.

[13] E. P. Wigner, Proc. Nat. Acad. Sci. **38** (1952) 449.

[14] H. Weyl, Z. Phys. (in German) **56** (1929) 330.

[15] H. Lehmann, K. Symanzik and W. Zimmermann, Nuovo Cimento **1** (1955) 1425.

[16] *Selected Papers on Quantum Electrodynamics*, edited by J. Schwinger (Dover, Mineola, NY, 1958).

[17] L. Morel, Z. Yao, P. Cladé and S. Guellati-Khélifa, Nature **588** (2020) 61–65.

[18] J. Schwinger, Phys. Rev. **73** (1948) 416–417.

[19] S. Laporta and E. Remiddi, Phys. Lett. B **379** (1996) 283–291, e-Print: 9602417 [hep-ph].

[20] T. Aoyama, T. Kinoshita and M. Nio, Atoms 7 (2019) 28.

[21] D. Hanneke, S. Fogwell Hoogerheide and G. Gabrielse, Phys. Rev. A83 (2011) 052122, e-Print: 1009.4831 [atom-ph].

[22] J. Chadwick, Nature **129** (1932) 312.

[23] W. Heisenberg, Z. Physik (in German) **77** (1932) 1–11.

[24] M. Gell-Mann, Phys. Rev. **125** (1962) 1067.

[25] Y. Ne'eman, Nucl. Phys. **26** (1961) 222–229.

[26] M. Gell-Mann, Phys. Rev. **125** (1962) 1067.

[27] E. Fermi, Ric. sci. prog. tec. econ. naz. (in Italian) **2** (1933).

[28] E. Fermi, Il Nuovo Cimento (in Italian) **11** (1934) 1–19.

[29] E. Fermi, Z. Physik (in German) **88** (1934) 161.

[30] C. N. Yang and T. D. Lee, Phys. Rev. **104** (1956) 254.

[31] C. S. Wu, E. Ambler, R. Hayward, D. D. Hoppes and R. P. Hudson, Phys. Rev. **105** (1957) 1413–1415.

[32] R. P. Feynman and M. Gell-Mann, Phys. Rev. **109** (1958) 193–198.

[33] R. Marshak and E. C. G. Sudarshan, Phys. Rev. **109** (1958) 1860–1862.

[34] S. S. Gershtein and J. B. Zeldovich, Zh. Eksperim. i Teor. Fis. **29** (1955) 698 [English transl.: Soviet Phys.—J. Exp. Theor. Phys. **2** (1955) 576].

[35] O. Klein, Nature **4101** (1948) 897.

[36] J. Bardeen, L. N. Cooper and J. R. Schrieffer, Phys. Rev. **106** (1957) 162–164.

[37] J. Bardeen, L. N. Cooper and J. R. Schrieffer, Phys. Rev. **108** (1957) 1175–1204.

[38] Y. Nambu, Phys. Rev. **117** (1960) 648–663.

[39] Y. Nambu and G. Jona-Lasinio, Phys. Rev. **122** (1961) 345–358; Phys. Rev. **124** (1961) 246–254.

[40] J. Schwinger, Ann. Phys. **2** (1957) 407.

[41] S. Glashow and M. Baker, Phys. Rev. **128** (1962) 2462.

[42] S. Coleman in *Aspects of Symmetry: Selected Erice Lectures*, Chapter 5 (Cambridge University Press, Cambridge, 1988).

[43] V. L. Ginzburg and L. D. Landau, Zh. Eksp. Teor. Fiz. (in Russian) **20** (1950), 1064 [English translation appears in L. D. Landau, *Collected Papers* (Pergamon Press, Oxford, 1965) p. 546].

[44] W. Heisenberg, Zeitschrift für Physik (in German) **49** (1928) 619–636.

[45] M. Gell-Mann and M. Levy, Il Nuovo Cimento **16** (1960) 705–726.

[46] J. Goldstone, Nuovo Cimento, **19** (1961) 154–164.

[47] J. Goldstone, A. Salam and S. Weinberg, Phys. Rev. **127** (1962) 965–970.

[48] C. N. Yang and R. L. Mills, Phys. Rev. **96** (1954) 191.

[49] M. Gell-Mann, lectures at the 1972 Schladming Winter School, CERN TH.1543 Acta Phys. Austriaca Suppl. (1972).

[50] H. Fritzsch and M. Gell-Mann in *Proceedings of the XVI International Conference on High Energy Physics*, edited by J. D. Jackson and A. Roberts (NAL, Batavia, IL, 1972).

[51] W. A. Bardeen, H. Fritzsch and M. Gell-Mann in *Scale and Conformal Invariance in Hadron Physics*, edited by R. Gatto (Wiley, New York, 1973).

[52] H. Fritzsch, M. Gell-Mann and H. Leutwyler, Phys. Lett. **B47** (1973) 365–368.

[53] O. W. Greenberg, Phys. Rev. Lett. **13** (1964) 598.

[54] Y. Nambu, in *Proceedings of the 2nd Coral Gables Conference on Symmetry Principles at High Energy*, 274–285, edited by B. Kursumoglu, A. Perlmutter, and I. Sanmar (Freeman, New York, 1965).

[55] Y. Nambu, in *Preludes in Theoretical Physics in Honor of V. F. Weisskopf*, 133–142 edited by A. De-Shalit, H. Feshbach, and L. Van Hove (North-Holland, Amsterdam, 1965).

[56] Y. Han and Y. Nambu, Phys. Rev. **139B** (1965) 1006.

[57] D. J. Gross and F. Wilczek, Phys. Rev. Lett. **30** (1973) 1343–1346.

[58] H. D. Politzer, Phys. Rev. Lett. **30** (1973) 1346–1349.

[59] V. S. Vanyashin and M. V. Terent'ev, J. Exp. Theor. Phys. **21** (1965) 375–380.

[60] I. B. Khriplovich, Sov. J. Nucl. Phys. **10** (1970) 235–242.

[61] G. 't Hooft, unpublished talk at the Marseille conference on renormalization of Yang–Mills fields and applications to particle physics (1972).

[62] P. A. Zyla, R. M. Barnett, J. Beringer et al. (Particle Data Group), Prog. Theor. Exp. Phys. **2020** (2020) 083C01.

[63] E. D. Bloom, D. H. Coward, H. C. DeStaebler et al., Phys. Rev. Lett. **23** (1969) 930–934.

[64] M. Breidenbach, J. I. Friedman, H. W. Kendall et al. Phys. Rev. Lett. **23** (1969) 935–939.

[65] J. I. Friedman and H. W. Kendall, Ann. Rev. Nucl. Par. Sci. **22** (1972) 203–254.

[66] J. D. Bjorken, Phys. Rev. **148** (1966) 1467; **179** (1969) 1547.

[67] R. P. Feynman, Phys. Rev. Lett. **23** (1969) 1415.

[68] R. P. Feynman, *Photon-Hadron Interactions* (Addison-Wesley, Boston, 1988).

[69] M. Gell-Mann, Phys. Lett. **8** (1964) 214–215.

[70] G. Zweig, CERN Report No.8182/TH.401; CERN Report No.8419/TH.412.

[71] J. D. Bjorken and E. A. Paschos, Phys. Rev. **185** (1969) 1975.

[72] C. Chang, K. Chen, D. Fox et al., Phys. Rev. Lett. **35** (1975) 9.

[73] R. K. Ellis, Adv. Ser. Direct. High Energy Phys. **26** (2016) 61–78.

[74] L. D. Landau, A. A. Abrikosov and I. M. Khalatnikov, Dokl. Akad. Nauk SSSR **95** (1954) 497.

[75] E. W. Kolb and M. S. Turner, *The Early Universe*, Frontiers in Physics **69** (Addison-Wesley, Boston, 1990).

[76] E. C. Marino, *Quantum Field Theory Approach to Condensed Matter Physics* (Cambridge University Press, Cambridge, 2017).

[77] K. Wilson, Phys. Rev. D **10** (1974) 2445.

[78] C. Gattringer and C. B. Lang, *Quantum Chromodynamics on the Lattice* (Springer, 2009).

[79] T. Degrand and C. DeTar, *Lattice Methods for Quantum Chromodynamics* (World Scientific, 2006).

[80] I. Montvay and G. Münster, *Quantum Fields on a Lattice* (Cambridge University Press, 1997).

[81] A. Kronfeld, Ann. Rev. Nucl. Part. Sci., **62** (2012) 265, e-Print: 1203.1204 [hep-lat].

[82] J. Schwinger, Phys. Rev. Phys. Rev. **128** (1962) 2425–2429.

[83] P. W. Anderson, Phys. Rev. **130** (1963) 439–442.

[84] W. Meissner and R. Ochsenfeld, Naturwissenschaften (in German) **21** (1933) 787–788.

[85] E. Stueckelberg, Helvetica Physica Acta **11** (1938) 299–312.

[86] G. Källén, Elementary Particle Physics (Addison Wesley, 1964).

[87] B. Pontecorvo, Phys. Rev. **72** (1947) 246.

[88] G. Puppi, Nuovo Cimento **5** (1948) 505.

[89] S. L. Glashow, Nucl. Phys. **2** (1961) 579.

[90] A. Salam and J. C. Ward, Novo Cim. **11** (1959) 568; **19** (1961) 165.

[91] S. L. Glashow, J. Iliopoulos and L. Maiani, Physical Review **D2** (1970) 1285.

[92] J. H. Christenson, J. Cronin, V. L. Fitch and R. Turlay, Phys. Rev. Lett. **13** (1964) 138.

[93] N. Cabibbo, Phys. Rev. Lett. **10** (1963) 531–533.

[94] J. J. M. Aubert, U. Becker, P. J. Biggs et al., Phys. Rev. Lett. **33** (1974) 1404–1406.

[95] J.-E. Augustin, A. M. Boyarski, M. Breidenbach, et al., Phys. Rev. Lett. **33** (1974) 1406–1408.

[96] M. Kobayashi and T. Maskawa, Progress of Theoretical Physics **49** (1973) 652–657.

[97] R. Brout and F. Englert, Phys. Rev. Lett., 13 (1964) 321.

[98] P. Higgs, Phys. Lett. 12 (1964) 132; Phys. Rev. Lett. **13** (1964) 508.

[99] G. Guralnik, C. R. Hagen and T. Kibble, Phys. Rev. Lett. 13 (1964) 585.

[100] A. Migdal and A. Polyakov, Sov. Phys. J. Exp. Theor. Phys. **24** (1967) 91–98, Zh. Eksp. Teor. Fiz. **51** (1966) 135–146.

[101] S. Weinberg, Phys. Rev. Lett. **19** (1967) 1264.

[102] A. Salam, in *Elementary Particle Theory*, ed. by N. Svartholm (Almquist and Wiksell, Stockholm 1968).

[103] P. Sikivie, L. Susskind, M. B. Voloshin and V. I. Zakharov, Nucl. Phys. **B173** (1980) 189.

[104] G. 't Hooft, Nucl. Phys. **B33** (1971) 173–177; Nucl. Phys. **B35** (1971) 167–188.

[105] M. J. G. Veltman, Nucl. Phys. B **7** (1968), 637–650.

[106] M. J. G. Veltman, Nucl. Phys. B **21** (1970), 288–302.

[107] G. 't Hooft and M. Veltman, Nucl. Phys. B **44** (1972) 189–219.

[108] G. 't Hooft and M. Veltman, DIAGRAMMAR, NATO Adv. Study Inst. Ser. B Phys. **4** (1974) 177–322.

[109] B. W. Lee, Phys. Rev. D **6** (1972) 1188; B. W. Lee and J. Zinn Justin, Phys. Rev. D **5** (1972) 3132, 3137, 3155.

[110] Gargamelle Collaboration, F. J. Hasert, H. Faissner, W. Krenz et al., Phys. Lett. **B46** (1973) 121.

[111] Gargamelle Collaboration, F. J. Hasert, S. Kabe, W. Krenz et al., Nucl. Phys. **B73** (1974) 1.

[112] Ya. B. Zel'dovich, J. Exp. Theor. Phys. **33** (1957) 1531, Ya. B. Zel'dovich, J. Exp. Theor. Phys. **2** (1959) 682.

[113] C. Y. Prescott, W. B Atwood, R. L. A. Cottrell et al., Phys. Lett. **77B** (1978) 347.

[114] C. Y. Prescott, W. B Atwood, R. L. A. Cottrell et al., Phys. Lett. **84B** (1979) 524.

[115] I. B. Khriplovich, *Parity Non-conservation in Atomic Phenomena* (Gordon and Breach, New York, 1991).

[116] M. A. Bouchiat and C. Bouchiat, Phys. Lett. **B48** (1974) 111.

[117] L. M. Barkov and M. S. Zolotorev, Pis'ma Zh. Eksp. Teor. Fiz. **27** (1978) 379 (in Russian), [English trans.: J. Exp. Theor. Phys. Lett. **27** (1978) 357].

[118] C. S. Wood, S. C. Bennett, D. Cho et al., Science **275** (1997) 1759.

[119] G. Arnison, A. Aniston, B. Aubert et al. (UA1 Collaboration), Phys. Lett. **B122** (1983) 103.

[120] G. Arnison et al. (UA1 Collaboration), Phys. Lett. **B126** (1983) 101.

[121] M. Banner, R. Battiston, Ph. Bloch et al. (UA2 Collaboration), Phys. Lett. **B122** (1983) 476.

[122] P. Bagnaia, M. Banner, R. Battiston et al. (UA2 Collaboration), Phys. Lett. **B129** (1983) 130.

[123] T. Aoyama, N. Asmussen, M. Benayoun et al., Phys. Rept. **887** (2020), e-Print: 2006.04822 [hep-ph].

[124] B. Abi, T. Albahri, S. Al-Kilani et al., (Muon g-2 Collaboration), Phys. Rev. Lett. **126** (2021) 141801, e-Print: 2104.03281 [hep-ex].

[125] T. Albahri, A. Anastasi, A. Anisenkov et al., (Muon g-2 Collaboration), Phys. Rev. D **103** (2021) 072002, e-Print: 2104.03247 [hep-ex].

[126] T. Albahri, A. Anastasi, K. Badgley et al., (Muon g-2 Collaboration), Phys. Rev. A (2021) **103** (2021) 042208, e-Print: 2104.03201 [hep-ex].

[127] G. W. Bennett, B. Bousquet, H. N. Brown et al., (Muon g-2 Collaboration), Phys. Rev. D **73** (2006) 072003, e-Print: 0602035 [hep-ex].

[128] Sz. Borsanyi, Z. Fodor, J. N. Guenther et al., Nature **593** (2021) 51, e-Print: 2002:12347 [hep-lat].

[129] ALEPH, DELPHI, L3, OPAL and LEP Electroweak Collaborations, Phys. Rept. **532** (2013) 119–244, e-Print: 1302.3415 [hep-ex].

[130] A. Denner, S. Dittmaier, M. Roth, D. Wackeroth, Nucl. Phys. **B560** (1999) 33–65, e-Print: 9904472 [hep-ph].

[131] S. Jadach, W. Placzek, M. Skrzypek, B. F. L. Ward, Z. Was et al., Comput. Phys. Commun. **140** (2001) 475–512, e-Print: 0104049 [hep-ph].

[132] D. Bardin, J. Biebel, D. Lehner et al., Comput. Phys. Commun. **104** (1997) 161–187.

[133] ALEPH, DELPHI, L3, OPAL, SLD Collaborations, LEP Electroweak Working Group, SLD Electroweak and Heavy Flavour Groups, Phys. Rep. **427** (2006) 257–454.

[134] S. L. Adler, Phys. Rev. **177** (1969) 2426.

[135] J. S. Bell and R. Jackiw, Nuovo Cim. **60A** (1969) 47.

[136] S. L. Adler and W. Bardeen, Phys. Rev. **182** (1969) 1517.

[137] W. Bardeen, Phys. Rev. **184** (1969) 1848.

[138] G. G. Ross, *Grand Unified Theories (Frontiers in Physics)* (Benjamin-Cummings Publishing Co., 1986).

[139] C. Abel, S. Afach, N. J. Ayres et al., Phys. Rev. Lett. **124** (2020) 081803, e-Print: 2001.11966 [hep-ex].

[140] W. A. Bardeen, Nucl. Phys. **B75** (1974) 246.

[141] R. D. Peccei and H. R. Quinn, Phys. Rev. Lett. **38** 1440–1443.

[142] R. D. Peccei and H. R. Quinn, Phys. Rev. D **16** (1977) 1791–1797.

[143] S. Weinberg, Phys. Rev. Lett. **40** (1978) 223–226.

[144] F. Wilczek, Phys. Rev. Lett. **40** (1978) 279–282.

[145] Super-Kamiokande Collaboration: Y. Fukuda, T. Hayakawa, E. Ichihara et al., Phys. Rev. Lett. **81** (1998) 1562–1567, e-Print: 9807003 [hep-ex].

[146] K. Nakamura, K. Hagiwara, K. Hikasa et al. (Particle Data Group), J. Phys. G **37** (2010) 075021.

[147] G. G. Raffelt, Lect. Notes Phys. **741** (2008) 51–71 (2008), e-Print: 0611350 [hep-ph].

[148] V. Anastassopoulos, S. Aune, K. Barth et al., Nature Physics **13** (2017) 584.

[149] J. H. Chang, R. Essig, and S. D. McDermott, J. High Energy Phys. **1218** (2018) 1.

[150] B. Pontecorvo, Zh. Eksp. Teor. Fiz. **33** (1957) 549–551; reproduced and translated in Sov. Phys. J. Exp. Theor. Phys. **6** (1957) 429–431.

[151] E. Majorana, Nuovo Cimento (in Italian) **14** (1937) 171–184.

[152] S. Weinberg, Phys. Rev. Lett. **43** (1979) 1566–1570.

[153] Z. Maki, M. Nakagawa and S. Sakata, **28** (1962) 870.

[154] P. Minkowski, Phys. Lett. B. **67** (1977) 421.

[155] T. Yanagida, Prog. Theor. Phys. **64** (1980) 1103–1105.

[156] J. G. de Swart, G. Bertone and J. van Dongen, Nat. Astron. **1** (2017) 0059.

[157] F. Zwicky, Helv. Phys. Acta **6** (1933) 110–127; Astrophys. J. **86** (1937) 217–246.

[158] J. H. Oort, Astrophys. J. **91** (1940) 273–306.

[159] V. C. Rubin and W. Ford Kent, Jr., Astrophys. J. **159** (1970) 379–403.

[160] V. Rubin, Science **220** (1983).

[161] WMAP Collaboration, D. N. Spergel, L. Verde, H. V. Peiris et al., Astrophys. J. Suppl. **148** (2003) 175–194, e-Print: 0302209 [astro-ph].

[162] G. Bertone, D. Hooper and J. Silk, Phys. Rept. **405** (2005) 279–390, e-Print: 0404175 [hep-ph].

[163] V. Trimble, Annu. Rev. of Astr. and Astro. **25** (1987) 425–472.

[164] D. Clowe, A. Gonzalez and M. Markevich, Astrophys. J. **604** (2004) 596–603, e-Print 0312273 [astro-ph].

[165] M. Markevitch, A. H. Gonzalez, D. Clowe et al., Astrophys. J. **606** (2004) 819–824, e-Print 0309303 [astro-ph].

[166] Sloan Digital Sky Survey, D. J. Eisenstein et al., Astron. J. **142** (2011) 72, e-Print 1101.1529 [astro-ph].

[167] D. H. Weinberg, R. Dave, N. Katz and J. A. Kollmeier in *The Emergence of Cosmic Structure: Thirteenth Astrophysics Conference*, S. H. Holt and C. S. Reynolds eds. (AIP Conference Series) **666** (2003) 157–169, e-Print 0301186 [astro-ph].

[168] M. Battaglieri, A. Belloni, A. Chou et al., contribution to *US Cosmic Visions: New Ideas in Dark Matter* (2017), e-Print: 1707.04591 [hep-ph].

[169] P. J. E. Peebles, Astron. J. **263** (1982) L1.

[170] G. R. Blumenthal, H. Pagels and J. R. Primack, Nature. **299** (1982) 37–38.

[171] P. J. E. Peebles and B. Ratra, Rev. Mod. Phys. **75** (2003) 559–606, e-Print: 0207347 [astro-ph].

[172] J. Frieman, M. Turner and D. Hutere, Ann. Rev. Astron. Astrophys. **46** (2008) 385–432, e-Print: 0803.0982 [astro-ph] (The term "dark energy" may first have appeared in the title of this paper.)

[173] A. G. Riess, A. V. Filippenko, P. Challis et al., Astron. J. **116** (1998) 1009–1038.

[174] Supernova Cosmology Project Collaboration: S. Perlmutter, G. Aldering, G. Goldhaber et al., Astrophys. J. **517** (1999) 565–586, e-Print: 9812133 [astro-ph].

[175] Supernova Search Team: A. G. Riess, L.-G. Strolger, J. Tonry et al., Astrophys. J. **607** (2004) 665–687, e-Print: 0402512 [astro-ph].

[176] G. Paál, I. Horváth and B. Lukács Astrophys. Space Sci. **191** (1992) 107–124.

[177] Planck Collaboration: N. Aghanim, Y. Akrami, F. Arroja et al., Astron. and Astrophys. **641** (2020) A6, e-Print: 1807.06205 [astro-ph.CO].

[178] Planck Collaboration: N. Aghanim, Y. Akrami, M. Ashdown et al., Astron. and Astrophys. **641** (2020) A6, e-Print: 1807.06209 [astro-ph.CO].

[179] A. Einstein, Sitzungsber. Preuss. Akad. Wiss. (in German), **142** (1931) 235–237.

[180] S. Weinberg, Rev. Mod. Phys. **61** (1989) 1–23.

[181] M. Dine and A. Kusenko, Rev. Mod. Phys. **76** (2003) 1, e-Print: 0303065 [hep-ph].

[182] A. D. Sakharov, J. Exp. Theor. Phys. Lett. **5** (1967) 32–35.

[183] G. t'Hooft, Phys. Rev. Lett. **37** (1976) 8; Phys. Rev. **D14** (1976) 3432.

[184] V. A. Kuzmin, V. A. Rubakov and M. E. Shaposhnikov, Phys. Lett., **B155** (1985) 36.

[185] V. A. Rubakov and M. E. Shaposhnikov, Usp. Fiz. Nauk, **166** (1996) 493–537.

[186] M. Fukugita and T. Yanagida, Phys.Lett.B **174** (1986) 45–47.

[187] W. Buchmuller, P. Di Bari and M. Plumacher, Ann. Phys. **315** (2005) 305–351, e-Print: hep-ph/0401240 [hep-ph].

[188] S. Davidson, E. Nardi and Y. Nir, Phys.Rept. **466** (2008) 105–177, e-Print: 0802.2962 [hep-ph].

[189] A. Andreassen, W. Frost and M. D. Schwartz, Phys. Rev. **D97** (2018) 056006.

[190] S. Chigusa, T. Moroi and Y. Shoji, Phys. Rev. Lett. **119** (2017) 211801.

[191] J. Elias-Miro, J. R. Espinosa, G. F. Giudice, H. M. Lee and A. Strumia J. High Energy Phys. **06** (2012) 031 e-Print: 1203.0237 [hep-ph].

[192] A. A. Starobinskii, J. Exp. Theor. Phys. **30** (1979) 682.

[193] A. Linde, Phys. Lett. B **108** (1982) 389–393.

[194] A. Linde, Lect. Notes Phys. **738** (2008) 1; V. Mukhanov, *Physical Foundation of Cosmology* (Cambridge University Press, 2005) 421.

[195] WMAP Collaboration D. N. Spergel, R. Bean, O. Doré et al., Astrophys. J. Suppl. **170** (2007) 377–408, e-Print: 0603449 [astro-ph].

[196] G. 't Hooft, NATO Sci. Ser. B **59** (1980) 135–157; Cargese Summer Inst. (1979) 135.

[197] J. Wess and H. Bagger, *Supersymmetry and Supergravity* (second edition). Princeton Series in Physics (Princeton University Press, 1992).

[198] W. A. Bardeen, in *Ontake Summer Institute on Particle Physics* (1995) 8, FERMILAB-CONF-95-391-T.

[199] J. E. Lidsey, A. R. Liddle, E. W. Kolb et al., Rev. Mod. Phys. **69** (1997) 373–410, e-Print: 9508078 [astro-ph].

[200] F. L. Bezrukov and M. Shaposhnikov, Phys. Lett. B **659** (2008) 703, e-Print: 0710.3755 [hep-th].

[201] A. De Simone, M. P. Hertzberg and F. Wilczek, Phys. Lett. B **678** (2009) 1–8, e-Print: 0812.4946 [hep-ph].

[202] F. Bezrukov, A. Magnin, M. Shaposhnikov and S. Sibiryakov, J. High Energy Phys. **01** (2011) 016, e-Print: 1008.5157 [hep-ph].

[203] J. L. F. Barbon and J. R. Espinosa, Phys. Rev. D **79** (2009) 081302, e-Print: 0903.0355 [hep-ph].

[204] D. Hilbert, *Nachrichten von der Gesellschaft der Wissenschaften zu Göttingen – Mathematisch-Physikalische Klasse* (in German) **3** (1915) 395–407.

[205] B. DeWitt, Phys. Rev. **160** (1967) 1113–1148.

[206] B. DeWitt, **162** (1967) 1195–1239.

[207] B. DeWitt, **162** (1967) 1239–1256.

[208] R. P. Feynman, Acta. Phys. Pol. **24** (1963) 697.

[209] J. F. Donoghue, Phys. Rev. D50 (1994) 3874–3888, e-Print: 9405057 [gr-qc].

[210] S. Carlip, Rep. Prog. Phys. **64** (2001) 885–942.

[211] M. Blau and S. Theisen, Gen. Relativ. Gravit. **41** (2009)743–755.

[212] M. Mangano and S. Parke, Phys. Rept. **200** (1991) 301–367, e-Print: 0509223 [hep-th].

[213] L. J. Dixon in *Proceedings of the Theoretical Advanced Study Institute in Elementary Particle Physics* (TASI 95) 539–584 (World Scientific, 1996, ed. D. E. Soper), e-Print: 9601359 [hep-ph].

[214] L. D. Faddeev and V. Popov, Phys. Lett. **25** (1967) 29.

[215] C. Becchi, A. Rouet and R. Stora, Phys. Lett. **B52** (1974) 344–346; Comm. in Math. Phys., 42 (1975) 127–162; Annals of Physics **98** (1976) 287–321.

[216] I. V. Tyutin, Lebedev Physics Institute preprint **39** (1975), e-Print: 0812.0580 [hep-th].

Acknowledgments

I thank Jesús Pérez-Ríos for initiating this endeavor and inviting me to contribute.

This contribution is dedicated to the memory of Robert D. Peccei; mentor, collaborator and good friend.

Cambridge Elements ☰

Physics beyond the Standard Model with Atomic and Molecular Systems

David Cassidy
University College London (UCL)

David Cassidy is Professor in Experimental Physics at University College London (UCL) where he leads the Positronium Spectroscopy group. His research is focused on positron trapping technology, particularly in the context of generating positronium atoms for use in optical and microwave spectroscopy and performing tests of fundamental physics.

Rouven Essig
Stony Brook University

Rouven Essig is Professor in Physics at the C. N Yang Institute for Theoretical Physics at Stony Brook University. His research focuses on searching for Dark Matter and other new particles beyond the Standard Model.

Jesús Pérez-Ríos
Stony Brook University

Jesús Pérez-Ríos is Assistant Professor at the Department of Physics and Astronomy of Stony Brook University. His research focuses on the study of fundamental processes in atomic and molecular systems, particularly as applied to chemical physics, condensed matter, and high energy physics.

About the Series

This Elements series presents novel atomic and molecular systems as a platform to study physics beyond the Standard Model, based on the synergy between high energy physics and atomic, molecular, and optical physics. The series covers several key areas of interest in this emerging field.

Cambridge Elements $\equiv$

Physics beyond the Standard Model with Atomic and Molecular Systems

Elements in the Series

Role of Symmetry in the Development of the Standard Model
Sherwin T. Love

A full series listing is available at: www.cambridge.org/PBSM

Printed in the United States
by Baker & Taylor Publisher Services

Printed in the United States
by Baker & Taylor Publisher Services